D1241738

I get by with a little help from my friends

It's not a "figleaf" of your imagination that the world looks like a different place the morning after an overnight thunderstorm. I love the landscape the morning after a thunderstorm complete with lightning. Fresh and green the atmospheric nitrogen wakes up a stuck in a rut world going through its paces and needing a good swift kick in the plants.

About to enter my sixties I recognize I missed my calling in life. In my childhood years I wanted to be an astronaut. In my high school years I wanted to be a journalist. Later I learned as an adult if only I could turn back the clock to be a meteorologist. This regret I carry with me and live out vicariously in my numerous meteorological friends. I would add a dash of magic dust to turn back the clock like the black magical inoculant we add for legumes like beans and peas. The beneficial micro-organisms attach to the roots of legumes to create swellings and absorb that beautiful atmospheric nitrogen before it becomes a forgotten weather statistic. It partners with the natural order environment and all is well.

The "lawn" and short of the story is almost 60 years have flown by on my journey. Instead of an astronaut, journalist or meteorologist, I became a "garden center retail business owner part-time media "sod"-caster and entremanure." Put that on your resume and try to explain it. Life truly does come at you fast. Love the plot you got.

"Rick's passion and ability for sharing his incredible knowledge of horticulture is a gift. He is a masterful storyteller with great wit and humor, planting seeds of inspirational lessons learned on the garden path of life. I'm confident Rick's book will be a fun and wild ride for gardeners of all ages."

— **Kerry Ann Mendez**
Award winning garden designer,
speaker and author *Perennially Yours LLC*

"If pun aficionado is a understatement, so is fine human. Rick is a character with character, and I'm proud to call him a friend."

— **Kyle Underwood**
WOOD TV 8 Meteorologist

"Rick Vuyst was ALWAYS my favorite television guest. Some of our best segments included "burning bush winged wahoo" … "wazoo" … and the best life advice disguised in gardening puns!"

— **Stephanie Webb, morning co-anchor**
Good Morning Suncoast, **ABC7 Sarasota Florida**

"It's a fact flowers make us happy, and digging in the garden relaxes us naturally. Rick Vuyst embodies the same. His mere presence makes one happy. With Rick you simply relax and settle in-naturally!"

— **J Schwanke**
The Flower Expert **on uBloom.com**

"As a beloved on-air personality in West Michigan for decades, Rick has always connected with his viewers and listeners through his unassuming nature, quick wit, expertise and never-ending puns. Whether in business or on a personal level, his energy, enthusiasm and zest for life should be an inspiration to us all!"

— **Lauren Stanton Morning Anchor**
and Host of *My West Michigan*
on WZZM 13 in Grand Rapids, MI.

"Rick has a passion that goes beyond his proclivities with plants: Rick's laughter, storytelling and wit are a gift to all of us who are planted in his proverbial garden. And, speaking from experience, his stories help us grow."

— Lisa Rose, Author
Midwest Foraging Herbalist,
Writer, Forager

ISBN 978-1-61808-168-1

Printed in the United States of America

Cover design created by Ron Bell of AdVision Design Group (www.advisiondesigngroup.com)

White Feather Press

Reaffirming Faith in God, Family, and Country!

I Just Wet

My Plants

Rick Vuyst

Table of Contents

Chapter One - Basil Instinct 1

Chapter Two - Wilma Plantsgrow 17

Chapter Three - My Mom told me there
would be daisies like this 33

Chapter Four - That is so Bosky 43

Chapter Five - Not Tonight Deer 55

Chapter Six - Operating by the seat of your plants 71

Chapter Seven - May the Forest be with You 91

Chapter Eight - Here comes Planta Claus 101

Chapter Nine - Lettuce Party like it's 1999 117

Chapter Ten - I just Wet my Plants 133

Chapter Eleven - Da Vine Intervention 143

Chapter Twelve - You are in for a Root awakening 153

Chapter Thirteen - Staying Grounded 163

Chapter Fourteen - Lawn and Order 171

Chapter Fifteen - How are your Gibberellins? 187

Chapter Sixteen - You can't go…. all the plants are
going to die 197

Chapter Seventeen - The Frozen Pundra 207

Epilogue - (Epic-Log) 225

Dedication

I dedicate the book to Hank Prins, who was firmly rooted and grounded in his time here on earth. Heaven received him early, because Hank was always ahead of me.

From the Author

It all started over a bowl of oatmeal. My friend Terry hounded me and my friend Hank Prins to do a radio show about gardening. It was the spring of 1993. Hank and I had turned Terry away numerous times, but he was the consummate salesman. Terry was in advertising sales for WOOD radio and wouldn't take no for an answer. He returned time after time, finally threatening to give the show slot to a competitor of ours.

We met for breakfast and it had to be oatmeal with no sugar and dry pancakes, because that's how Hank rolled. He was health conscious and rode his bike everywhere year round. Even a snow storm would not stop Hank, as you would see him biking through snow in January. We agreed to do the show, and the debut would be the first weekend in May of 1993 with Phil Tower.

Hank and I had been friends for years. Terry and I became friends as we developed the radio show. We had great fun together on vacations in Mexico and as the Blues Brothers, performing with a band at a resort in northern Michigan. Hank and I grew the show, and it quickly expanded from the initial one-hour offering to a two-hour show.

Hank developed headaches in the winter and spring of 1996, and had an eventual diagnosis of brain cancer that spring. By June 1st he had passed away ... far too young for his time on earth to be done. I was in denial this could happen, and it was difficult to wrap my mind around it. Hank had three sons, and my family and his family bonded as my kids and his grew up together.

After taking some time off from the show, I decided to continue the Flowerland show, and, with the support of family, friends and our listeners, the show continues to grow to this day. Doug and Kristi joined the show after Hank's death, and have been dedicated partners in its development. For a period of time after Hank's death, my daughter Angie was the program director and engineer of the show. Chapter 16 is dedicated to the energy and spirit she brought to the program.

As I write in Chapter 10 of this book, with plants as in life, some people leave us or die after blooming. Some are given the opportunity to continually reinvent themselves. That is why sharing our stories is so important. There is what I consider a beautiful new word out there called "sonder" which means adopting "the realization that each random passerby is living a life as vivid and as complex as your own." I dedicate the book to Hank Prins, who was firmly rooted and grounded in his time here on earth. Heaven received him early, because he was always ahead of me. Hank was not given the opportunity to reinvent himself, but I was, so I've been "reinventing" and repairing myself ever since. Hank was always willing to call me out if he felt I was making a mistake or an error in my life. I can remember clearly each time he did it, and its why, as people courteously ask "how are you" when calling into the show I say, "staying grounded."

I thank my family, Sue, for all the sacrificed Saturdays, Chris, Clara, Angie, Rick, Jenn, Max, Stacey, Jon, Ollie, Miggy, Micah, sister and brothers and others who have supported and encouraged me in this journey. I thank my business partners,

Sid for believing in me at a young age, and all the countless others in the Flowerland family that I have had the privilege to work with, both now, as well as those who have moved on. I thank all my media friends, too numerous to mention, but all gave me a kick in the plants and encouragement in my journey. I thank my friends, too numerous to mention, for your support and encouragement. I thank Abigail for the inspirations and friends from Lola to Audrey to Cal for your puns. I thank the listeners and viewers through the years who have encouraged me in my work. The highest compliment is from those who say they don't garden or don't own a home, but love to watch and listen anyway. I thank my friends in the GCPG for their mentoring, encouragement and support. I thank my drama coach from years ago who gave me the chance to solo on a brightly lit stage. The bright lights blinded my eyesight to the audience, yet increased my awareness of their presence, as I learned to feed on their energy, laughter and response. I wanted to do better each time out. To my ophthalmologist who mysteriously passed away shortly after freeing me from the glasses I had worn since a kid, giving me my 20/20 vision to read and write. Finally I want to thank my German soccer coach from high school who made me captain of a team that went 0-13 with one tie. It taught me at a young age to give up on sports and that I would never make a living from professional sports. You have to lose before you can learn how to win. I thank Skip Coryell for being a sounding board and for reminding me "I always exist one step removed from being either brilliant or a dumpster fire."

Both Hank and Sue would remind me that, when I thought I was running away from demons, I was instead running towards something better. There is, my friends, something better in this life. As my friend Dr. Allan Armitage would say when interviewing him on the show, "You can't grow old when you're looking forward to the future. Whether gardening or living life, you might have a sore back and an empty wallet, but make it fun." That is what this book is all about.

– Rick Vuyst

There was a young man straightened out
When told to eat veggies would pout
With many appeals
He would force down his meals
And soon he began to sprout

Chapter One

Basil Instinct

(The making of an Entre-Manure)

IT WAS A HOT SUMMER DAY. THE BAKING sun and wind as well as humidity left both man and plants parched for refreshment. As a young boy I sought refuge in the water from a garden hose attached to the house. It was hot outside and I had a drinking problem. Yes, I drank Tang, lots of it. First introduced in 1957 a couple of years before I was born it was popularized by an American hero John Glenn who allegedly drank it on a NASA Mercury mission. We were from the Apollo age where Tang was the choice of astronauts, and if it was good enough for them it was good enough for us. If our astronaut heroes drank Tang, we drank it while gazing at the moon, anticipating someday we would walk on the moon.

I had a heightened imagination as a kid, acting out heroic superpowers long before video games and the internet were on our radar. The same scientist who invented Tang also invented Cool Whip. I love Cool Whip, and the containers have come in handy as makeshift Tupperware containers for years. It's something you just can't throw away. They are perfect for storing nuts and bolts and other stuff you'll never use. I would eat Cool Whip right out of the container, and sometimes we would even mix the powdered fruit of Tang in water, taking it straight on the tongue, giving us space-age powers. Our lips and tongues would glow like a radioactive planet but we didn't care. We were living the dream.

When I was 11 years old Astronaut Stuart Roosa, a former U.S. Forest Service smokejumper, carried about 500 tree seeds into space as part of the three-man Apollo 14 crew that went to the moon in early 1971. Upon their return to Earth, the seeds were germinated, and most of the seedlings were given away to be planted as part of 1976 bicentennial celebrations across the country. While commander Alan Shepard was hitting golf balls in space, command module pilot Stuart Roosa was carrying future generations of tree seeds into orbit.

One such tree, with the help of local native President Gerald Ford, found its way to West Michigan. Unfortunately, the tree in later years was mistakenly cut down in a landscaping mishap. In 2009 I partnered with Wyoming police Officer Adam Bartone to bring another descendant of Roosa's sycamore trees back to West Michigan. The event was complete with a ceremony attended by NASA and US Forest Service representatives and other dignitaries. Together, we planted an offspring of a first-generation moon tree which today stands on the grounds of the Wyoming Police department. It's the closest I'll get to space short of gazing at the moon and stars from the beach or watching *Apollo 13* for the 107th time.

I have been told I am a goal-oriented, driven person, who moves quickly when there is a job to be done. Without the aid of a couch session with a psychologist taking me back to my childhood years, I recognize the foundation or launching pad of these characteristics. I was born and raised in the Apollo moonshot dreamers generation, when you instinctively learned that, just like the space program, three things were needed to get something done:

- I would need a goal
- I would need a plan (made up as I go)
- And not enough time to do it

When I wasn't drinking Tang and fantasizing about being an astronaut, we used packets of Funny Face powdered drink mix in tap water and Nestle Quik powder in our milk as other drinks of choice. But on a hot summer day, while playing outside, the drink of convenience was a shot of water from the garden hose. It tasted funny but was the refreshing water choice long before juice boxes and bottled water. Fast forward to years later and I reflect on those hot summer days as a kid. Today there are warnings not to drink from the garden hose because it isn't manufactured to deliver safe drinking water. I wondered why I should care when as a kid we never wore bike helmets when riding our bikes or seat belts for that matter in my Dad's 1955 Chevrolet. We would lie in the window of the back seat and laugh at being catapulted to the front of the car when braking. Today I read that those gulps of water from the garden hose, baking in the summer sun, have bacteria, mold and toxic chemicals ranging from lead, bromine, phthalates and BPA to toxic vinyl chloride which according to scientists today can lower intelligence, damage the endocrine system and cause behavioral changes. As far as I'm concerned it explains a lot. I have often thought of bottling warm water that tastes like summer time garden hose water. I'm sure it would sell as a novelty to all those who, like me, remember what it was like to spend summers outside internet free, bored and adventurous as a kid in search of refreshment from a garden hose.

I remember walks with my Dad on Sunday afternoons. My Dad has a fertile mind just like me. We would walk together through fields on a Sunday afternoon and he always made it a point to bring me to some grasses in an area of the field that appeared to have teeth marks in the foliage. He would carefully explain to me that the teeth marks were a curse, a continual curse of the frustrations of Adam and Eve in the garden. When banished from the Garden of Eden their imprint on these grasses remain today symbolic of their frustration and frailty, a living legacy and lesson to all who, like me, face their imperfections and mistakes later in life. As a young boy I bought the concept hook, line and sinker. He was my Dad, and he knew everything and was never wrong. As a grown man I understood he had stretched the truth without scientific evidence, but had the liberty to do so with an impressionable boy. As an adult I learned the grass was an invasive species called *Phragmites* or common reed, a monoculture that pushes out diversity in marshes and wetland areas. I became familiar with this alarming invasive species like Asian carp, emerald ash borer, purple loosestrife or woolly adelgid and how they dramatically change the landscape we live in. In the case of *Phragmites australis*, we became all too familiar with it living on the shore of the Great Lakes. In wet, stormy rainy seasons some would dislodge and float down the rivers emptying into Lake Michigan. The vegetation would then roil and churn in the waves and water collecting trash as it went, eventually finding its way to shore. Lining the shore, it would then cook in the hot, July sun creating an awful stench and less than inviting unsightly conditions for beach goers. In the case of the invasive species, common reed, it has notches in the foliage that with wild imagination could be seen as teeth marks. It didn't matter, my Dad had made an impression on my young, fertile mind that sticks today.

When it came to being groundbreakingly hip, my Dad would compost in the garden in fall when everyone else was burning leaves curbside. I remember the eerie, smoky haze throughout that old neighborhood in late October as neighbors would rake leaves into the street to burn them against the curb. Piles of smoldering leaves looked like the wreckage of some urban battlefield conflict as we navigated our banana-peel bikes through the acrid and dusky streets. The smoke would swirl, blocking the sun at times and burning your eyes. Evenings spent navigating those streets had to be equivalent to smoking a pack or two of non-filter cigarettes. One late afternoon, while on such a reconnaissance mission to obtain candy cigarettes from the drug store, some friends and I reconnoitered our position from the safety of the sidewalk planning our next move. We evaluated the speed at which our bikes would have to travel to make it through the smoldering foliage of street-side leaf piles without incident. Our mission's success would depend on the speed of our craft and whether Mom would find out later due to some unforeseen malfunction. I volunteered to serve as cavalry scout and make the first attempt while the third-grade platoon observed from a safe distance.

Accelerating to top speed, I hit the first pile of brown foliage and tore through the smoldering cauldron of oak leaves. I watched the swirl of leaves and the belch of hot gases and smoke behind me, pleased my first foray was a success. The second pile however did not work out as planned. Being a much larger pile, my bike quickly slowed and pinned to the curb. Sensing the heated quagmire I had put myself in I knew I had to abort the mission. I had lost momentum and found myself stepping off into the glowing magma of autumn embers and angry, inflamed foliage. A mountainous volcano of hot lava quickly melted the tires of my bike and singed my ankles. My band of brothers quickly dispersed to their home bases, and I was left to walk my bike home with flat, melted tires. Covered in soot, with burnt tennis shoes and wrecked transportation, I looked and smelled like I had battled a forest fire that afternoon. To this day I have a red badge of courage on my ankle, commemorating the mission that crashed and burned that autumn afternoon. The state of Michigan outlawed curbside burning after the '60s, sparing future pioneers as myself further "watch this" dares that young boys make in naive bravado.

To this day, I think back to those burning piles of leaves whenever the topic of cutting back ornamental grasses comes up on the radio show. Kristi, the "cracked pot" of two fun-guys and a cracked pot fame, loves to brag that her method of cutting back dead ornamental grass foliage tops all others. Once the foliage of a spent *Miscanthus* is brown and crispy, she proceeds in lighting up the remnants with the use of a lighter and a can of Aqua Net hairspray. As I understand it, aerosol cans were developed by the Department of Agriculture during World War II to better distribute insect spray to soldiers and deal with the prevention of malaria. After the war, the beauty industry saw an opportunity, and hairsprays like Aqua Net became a big seller with women who wanted the all-day hold of the masterpiece on their head. My dad called them bird nests, and I remember thinking what self-respecting bird would want to live in that toxic, high-rise housing project. I recall observing ladies in the '60s with their bouffant and beehive hairdos, wondering how they slept at night. I now understand that those early versions of hairspray contained the propellant vinyl chloride. I'm no rocket scientist, but those ladies in church when I was a kid were wearing a sticky mountain of ozone-busting explosive toxins on their heads. No wonder the peppermints they distributed during eternal sermons about eternity tasted so funny. The white and pink orbs of sugary stillness tasted like perfume with a hint of aerosol propellant. They had no grounds to critique my leaf-pile adventures when it came to sustainable, environmentally friendly activities and were not setting a good example to a young, moldable mind such as mine. I'm amazed at how these childhood memories stick with me and how ornamental grasses today can bring these recollections to the surface.

Kristi lives out in the middle of nowhere so she can burn her grasses with a little help from Aqua Net. I however do not recommend the method especially in urban neighborhoods. Watch a dry six-foot ornamental grass combust into flames on a YouTube video and you'll know what I mean. I looked at the can Kristi gave me for my birthday one year. I keep it on a shelf in my office. I can live with the Aqua (water) part of the product. It's the "net" part of Dimethyl Ether, Vinyl Neodecanoate Copolymer Acrylates and Aminomethyl Propanol, along with other ingredients, that give me pause as to why I have it on the shelf next to my desk. I have the super-hold product in a lovely, pink can that features a "fresh fragrance." I plan to use it at a future retirement party. Until then it will serve as a retro paper weight.

Much to my chagrin my parents were fashionably current, understanding it was hip to eat kale before it became hip to eat kale. The Dutch enjoy a comfort food called Boerenkool or Stamppot, also known as kale hash. I am all for healthy and I like kale in my salads today. But as a kid, warm kale mashed into lumpy potatoes was pure torture. No one should be forced to eat kale, especially warm kale mashed into potatoes garnished with greasy sausage. The rookworst, appropriately named because it was one of the worst things I've ever eaten, is a type of Dutch sausage. It looks like a bologna-type sausage, and uses natural intestine casings or bovine collagen to hold it together. Upon breeching the casing of the sausage with your teeth, the greasy cholesterol rushes to your bloodstream where it resides, I surmise, in my arteries today.

In the evening, I would open the back door and enter the home, praying the unmistakable aroma of this Dutch delicacy would not hit my nostrils. The sound of sizzling rookworst and the disheartening smell of boiling kale in the kitchen would crush my spirit. All hope was lost. It was going to be a long night. Looking back, I realize it was my Dad's job to reinforce my Mom's position that "she made it so I'd better eat it." Upon one such skirmish at the dinner table, I told my Dad I would rather eat horse manure and grass than eat this revolting fare. My Dad's response was he would take me up on that challenge. It was his intention to quell this adolescent uprising. Boerenkool battle lines were drawn. The battle of Boerenkool was inevitable and the conflict would be historic. A line in the sand had been crafted, and I was not about to move. Entrenched in my position,

I reasoned it was better to die there at the table and champion my cause than compromise my principles. I would go to bed hungry and consider picketing the sidewalk in front of the house the next day. In the '60s everyone on TV was walking around with picket signs championing their position. I was making a sacrifice for Dutch kids all around the world.

The one redeeming characteristic of Dutch fare, however, was the morning after the Boerenkool war. The Dutch have a peculiar culinary habit of eating chocolate for breakfast and kale for dinner. We would eat something called hagelslag, which is the American equivalent to the sprinkles we put on sundaes at the ice cream sundae bar. Making toast slathered in butter and topped off with chocolate sprinkles or multi-colored candy fla-vored sprinkles was perfectly normal. As the smoke in the kitch-en cleared from the Rookworst battle and the great Boorenkool war, a ceasefire was declared and all was well on the home front as we dined on our sugar-coated toast. What I really wanted was Boerenjongens, a Dutch dessert of raisins, sugar and alcohol. It wasn't until later in life when this became a hot topic on the radio show that I learned my parents were holding out on me.

I really had no "grounds" to fight my Dad when it came to eating kale or any other vegetable for that matter. It wasn't un-til later in life that I realized my Dad and Mom survived the "Hunger Winter" of 1944. The Dutch famine in the winter of 1944/45 was caused by the falling apart of the Nazi war machine. Although some of Europe had been liberated, the Netherlands remained Nazi-occupied, and the German occupiers stripped the country of food and fuel for use by Germany. During a harsh winter with little food and no fuel, Dutch citizens, including my Mom and Dad, were forced to fight for survival. My Dad would have been 15 at this time and my Mom about nine years old.

Real hunger makes you eat everything you can get, including tulip bulbs. Tulip bulbs plentiful and unplanted due to the war. During the famine the dry tulip bulbs were an option to keep

Dutch citizens, including my parents, alive. Recipes were created using the dry tulip bulbs originally intended for the ground and spring blooms. My Dad's family ate tulip bulb soup. With water, an onion and some cut-up tulip bulbs and salt, tulip bulb soup became a way of life. I have heard descriptions of tulip bulbs from "mealy" to "milky sweet" to "bitter and starchy" tasting something like wet sawdust. My Dad said they were better than the rations of sauerkraut mixed with sugar beets. The key was to remove the outer skin called a turban and to slice them, removing the center yellow core which was both bitter and could cause serious digestion problems making you sick. I've eaten tulip bulbs. They are awful. Not as bad as Booerenkool but distasteful none the less. It helped me understand that I do not, have not, and in no way understand true hunger.

Who would think that years later I would be selling tulip bulbs for a living in my store? My parents had to store and eat tulip bulbs to live. The irony of this story helps ground me. You do what you need to do. The Bedouin tribes of the Middle East, as I understand it, would eat tulip bulbs raw. I suppose if you're in the middle of the desert riding a camel a tulip bulb could be a handy snack. I still can't envision myself eating tulip bulbs and am doubtful it would "grow" on me.

ONE SATURDAY MORNING ON MY RADIO SHOW, I BROACHED a subject from my childhood memory, and was shocked to find it was foreign to others. I used to wear bread bags in my boots. In winter it was customary before putting on boots to first slip bread bags over my socks held up with rubber bands. My favorite brand was Wonder Bread and I thought everybody did it. You tried to make Wonder Bread bags last because they were much better than the generic bread bags. Thinking back, the bread bags were more useful than the white bread inside them, that would stick to the roof of your mouth well after lunch period. Moms in the 60's had a bread bag drawer, and, if you were lucky, she turned them inside out, shaking out the bread crumbs before you put them on.

Upon mentioning the topic on-air, you would think I had just landed a space ship from a distant planet and emerged with vegetation growing out of my ears. It was ingrained in my memory, and I assumed, as kids, that we all wore bread bags in our boots to keep our feet dry. The topic lit up the phone lines and we were on a "roll." I did find some support and encouragement from others who had the same experience. The topic was good for a "loaf," and I'm reminded often of my fashion faux pas in elementary school. Little did I know that the '70s, with leisure suits, silk shirts, long hair and white elevator shoes, was waiting right around the corner for this champion of haute couture. Man shall not live by bread alone, so be resourceful and wear the bread bags.

I was always outside as a kid ... winter, spring, summer and fall. I am a firm believer in getting kids outside to exercise their discovery tendencies and to get dirty. Even as an adult, dirt has proven, natural anti-depressant qualities. It is natural to have dirty thoughts, so go dig up some soil and get yourself grounded. Long days outside caused us to make up activities to pass the time. A favorite activity was using old, spent shovel handles as a baseball bat to hit projectiles, namely rocks, into the field. We would spend hand-numbing, blister-causing hours whacking stones with a shovel handle, pretending to hit a game-winning home run in the ninth inning. Beneficiaries of plaudits rained down upon us in our minds for our game-saving heroics, and hours were spent collecting buckets of rocks perfect for our next batting practice. The activity got juicy when we graduated to hitting green walnuts or osage oranges. There was something intriguing about the oddly shaped, bumpy, lime-green fruits of osage orange that looked like the brains of small space aliens on earth. They made perfect batting practice softballs, and we would head home with the green, fleshy debris of their carnage stuck to our clothes and faces.

Maclura pomifera or osage orange is a deciduous shrubby tree with an interesting history. It was highly regarded by Native Americans, who would use the durable and pliable but very strong wood to make bows. The wood of the osage orange was highly prized, and would prove valuable in trade for blankets and other traded goods. French traders would receive young "osage apple" plants or slips from the Osage Indians, and one of those traders passed some along to Meriwether Lewis on a Lewis and Clark expedition. Lewis passed along cuttings to Thomas Jefferson and wrote of its value to Native Americans. This discovery west of the Mississippi was soon planted at Monticello and other areas in the east. I recall seeing osage orange in visits to Monticello, and I believe I saw it along the fence that surrounds the grave site of Thomas Jefferson in Virginia. Esteemed at the time by Native Americans for bows, fiber for rope, tool handles or dye, it was used by settlers for wagon wheels and hubs, as well as thorny fence rows to delineate property lines. Osage orange wood poles held up telegraph lines in the 1800s and in the early 1900s its dye was used to color the olive drab uniforms of World War I soldiers. Today fencing of course has evolved and there is little value for an osage orange as a landscape plant. The blemished and distorted fruit bombing your car in the driveway in autumn would make an insurance agent cringe. But a disfigured osage orange fruit in the hands of a shovel handle wielding boy in October? Priceless.

> Man shall not live by bread alone, so be resourceful and wear the bread bags.

THE MOMENT

WE ALL HAVE THAT MOMENT OR MOMENTS IN OUR YOUNGer years that we can point to as a turning point … something we carry with us the balance of our life, bound to come out in a counseling session with a psychologist. (One of those "and how does that make you feel?" moments.) Long before our older years, when we naturally develop a hearing aid that filters out criticism and amplifies compliments, we have those seemingly insignificant scars. It brings up the point, how many psychologists does it take to get a hydrangea to change its color? The answer is one. The hydrangea has to want to change. Like a good trellis you need a support group to get this stuff out. All those childhood memories are entrenched in your mind. Consider that the '60s of my childhood years, before entering my teens, would end watching grainy images on a black-and-white TV as Neil Armstrong and Buzz Aldrin took one small step for man and one giant leap for mankind. A significant, impressionable moment for a soon-to-be teen. My impressionable life-changing moment, however, involved a confrontation with my older sister. Years beyond that defining moment I wanted change for the good … to grow from that point forward. Fortunately, years later, I cleared the air with my sister long before professional therapeutic intervention was needed.

I had stepped out in the cool of the morning to cultivate my garden. Having nurtured my little piece of earth, I was determined the vegetables would grow to fruition under my watch. I had no lease on this land, no ownership, just the benevolent gift of space from my Dad to try my hand at gardening. The garden was positioned behind the garage, out of sight, but certainly not out of mind. I was on a mission to grow, not to necessarily eat the fruits of my labor. Pop Tarts were more my speed at that age than broccoli, but this was to be one of those growing experiences. Pop Tarts, when initially introduced in the mid '60s, were not frosted. By 1967 it was found Pop Tarts could be frosted and survive the toaster. This was better than landing a man on the moon! It was like having cake sanctioned and approved for breakfast. It was wonderful being 8 years old in the Apollo age of discovery, and breakfast would never be the same. Two frosted Pop Tarts from the toaster washed down with a glass of Tang to start my day. Life was good.

It was in one of these sugar-euphoric moments I would begin the day by inspecting my garden. Captain of my earthen vessel I would cultivate their growth with encouraging care. Turning the corner behind the garage my mouth hung open, and I couldn't believe my eyes. All of my plants were uprooted and laying on their sides, gasping for life. Most had already left the land of the living, wilted beyond hope, having moved on to the great compost pile of vegetation. Who would do such a thing? Blatant and cruel, an involuntary slaughter of the plants I had nurtured from seed.

I reported the cruel caper to my Mom and Dad. I had one solitary suspect based on a theory I had developed. If my plants were pulled and laying on their side, but my sister's plants were still intact and growing fine right next to them, the mere coincidence would tip the scales of justice. Any logical jury would side with the prosecution that she had the sole motivation to commit such a crime. Her curiosity to see what was going on below the earth's surface had been more than she could take. My sister had succumbed to the temptation and committed involuntary plant slaughter.

Soon the pressure became more than she could handle. The walls were closing in on her and she cracked. A note was left for me with guised confession from the lead suspect in the case all along. My sister, knowing she could not escape justice and in frustration, penned the following note to me:

> *Dear Rick,*
>
> *I hate you.*
>
> *Love, Jane.*

A veiled confession … and a horticulturist was born. I would seek revenge years later by becoming a famous Entre-Manure.

A ROCKY ROAD

IT WAS A SUNNY, BRIGHT DAY AFTER A MARCH SNOWSTORM in Philadelphia, and much of the snow from the previous day had already melted. Having attended the Philadelphia Flower show, I decided it was time to put on my running shoes, inspired to run the streets of Philadelphia like Rocky Balboa. I left the hotel and ran to Independence Hall where I would begin my 2.5-mile sprint to the iconic steps of the Philadelphia Museum of Art and then back.

I ran from Chestnut Street to Arch Street, sensing the American history that just oozes from the surrounding environment. From Arch street I ran through Chinatown, mimicking Sylvester Stallone, past the bustle of vendors selling produce and groceries, waving at those who would glance my way as I went on to Vine street.

I was in high school when the first Rocky movie came out in 1976, and I remembered how I was inspired by his training runs. After watching the movie as a 17-year old, I donned sweat pants and got a jump rope. I wasn't willing however to get up at four in the morning and drink raw eggs from a glass. I had to draw the line somewhere.

Fast forward, as I ran through the streets of Philadelphia, now a grown 57-year-old man, I shadow boxed, jabbing the cool March air as I dodged garbage cans and light poles, traffic and cross streets as if in my own cinematic episode as "Ricky" Balboa.

I ran down Vine Street to Logan square, and then up the Benjamin Franklin Parkway, past The Thinker and the Gates of Hell at the Rodin museum. Thinking I should not abandon all hope as the steps of the Art Museum came into view, I crossed the parkway, saluting the impressive Washington Monument. I was on the home stretch. There, in front of me, at last stood the numerous stone steps to the top for my own personal Rocky moment.

I was far from alone. On a sunny day the steps are a tourist draw, and it's fun to watch from the top the human struggle below to get there. People, determined to get to the top, will take breaks along the way and sit on the steps sipping on their water bottles to catch their breath. I ran to the top and turned raising my arms, admiring the impressive view of the Philadelphia skyline and parkway. For a high school boy in the '70s this was truly a Hollywood moment.

I ran back to Market Street past the ornate Philadelphia City Hall. Running east on Market street towards the Delaware River and feeling quite proud of myself … it happened in an instant. I went down unceremoniously, like a sack of potatoes, landing on the pavement before I knew what had hit me. I pulled myself up to a sitting position as traffic sped by. Looking back, I saw a metal ring lying on the walk, and realized I had inadvertently tangled both feet inside the ring sending me down in an instant. As is human nature, I remember being more concerned with my embarrassment than the nature of my injuries. I popped to my feet and continued back towards my hotel. Entering the lobby with torn jacket and pants, bleeding from one knee and an arm, I looked like I had gone a round with Apollo Creed.

Just like Vince Lombardi said, "It's not whether or not you get knocked down; it's whether you get up." Things are going to get a little rocky now and then. It's only natural.

Spring always follows winter, and, as I like to say, "Setbacks add seasoning to better days ahead."

Usually, when you're cruising along and proud of yourself, something is going to knock you back a couple pegs, and a rocky road will set you straight again. When winter days are grey, and it's dark and cold, just remember it's only "adding seasoning to better days ahead." They will only be that much sweeter as together we head down the rocky road.

Watering plants until they drown
She was in for a big let down
Her skills without practice
Could kill a live cactus
Her thumb was perpetually brown

Chapter Two

Wilma Plantsgrow

I USE A MICROPHONE AS MY PREFERRED gardening tool to cultivate results. It's all theatre of the mind, and the end goal is to help people relax, smile and maintain their "composture." It's about giving people confidence, using 50,000 watts of effective radiated power on the FM side and 20,000 watts of power on the AM side, with a side dish of on-line streaming and podcasts for dessert. The show is like anything else in life, if you work hard and are kind, good things will happen. I like to envision the microphone if used properly as an effective and powerful clod-busting, weed smothering, planting, digging, soil-amending tool that you won't find for sale on the shelves of your local DIY home-improvement store.

I had the opportunity to sit in for Ralph Snodsmith host of the 'Garden Hotline' on WOR radio in New York one spring day in 2010. Mr. Snodsmith was a gardening legend who broadcasted on WOR "the voice of New York" for 35 years. His show was broadcast on more than 100 stations around the nation. His book *Tips from the Garden Hotline,* written in 1984, was a paperback purchase I made at the time and I wore the book out. Like an understudy, I analyzed his ability to work with callers to help them get their Christmas cactus to bloom or turn an avocado pit into a plant using toothpicks and a glass of water.

I had a great time doing the show, knowing all the while my midwestern accent was unusual to them. I don't have the stereotypical distinctive manner that my friends on the east coast have. When my New York friends talk, words like park sound like "pak." I was a Midwesterner doing a "tawk" radio show in New York. Three sounds like "tree" and we enjoy walking along the "watta" of the Hudson "rivva." Even my daughter Angie, who has lived in New York city now for years, has picked up the habit of dropping Rs if they are before a consonant or at the end of a word. A caller named Edith from Brooklyn was enamored by my accent as though I was from a foreign country. My everyday accent was distinctive to their ears on the radio. Being from the Midwest I understand that when someone says, "you have a distinctive sound" it is their passive-aggressive way of saying, "you sound funny."

The show went great and the two hours flew by. The following day the program director thanked me for doing the show, and his only critique was that I was "too nice" ... something they call "Midwestern nice." Not passive-aggressive, just the fact I used the words thank you and please and wasn't critical of a caller's Calathea. Thank you very mulch.

Instead of nice I thought they meant Midwestern "ice" as in the unremitting cold and snow we experience for four months here in the Midwest. Living on the Lake Michigan shoreline, when the arctic clippers come out of the northwest the snow piles up like it does in Buffalo, New York or Erie, Pennsylvania. Just like people in the southwest say, "yeah but it's a dry heat," we in the Midwest say, "yeah, well at least we get the four seasons."

After being pounded with snow and cold for four months, the spring season is quite amazing here in Michigan. The *Galanthus* snowdrops and crocus pop their heads out of the soil, and the *Hamamelis vernalis* or Witch Hazel come into bloom in very early spring with their curled and crinkled bright yellow flower petals. Hamamelis saves the best for last, however, with an amazing show of color in fall. It has an interesting history as the branches years ago were used for dowsing or as divining rods to locate the presence of water or metals. I use it to locate the presence of my yard after 100 inches of snow in winter. It, like *Viburnum*, *Fothergilla* or maple tree, is a symbol of the four seasons experience we enjoy as gardeners here in the Midwest.

The evening before doing the show on WOR, I remember lying in bed awake thinking, *I'm sure the first call I get will be from one of those cities where I stumble over the pronunciation.* It may sound silly to think of something like that, but people take their locale integrity personally. Anyone in media will tell you, just pronounce Oregon wrong once and people will come out of the woodwork correcting you. My mind circled and I mentally practiced the pronunciations of what I knew was to come.

> Being from the Midwest I understand that when someone says, "you have a distinctive sound" it is their passive-aggressive way of saying, "you sound funny."

Aquebogue New York

Patchogue New York

La Jolla California

Skaneateles New York

Worcester Massachusetts

Poughkeepsie New York

Puyallup Washington

I've been to Puyallup a number of times and still mangle the pronunciation. My friend Maidee who lives there has tried for years to help me through it, but invariably I bungle the attempt. Instead of saying "pew-al-up" I say something like "pile-up" as in a fender-bender on Interstate 5.

The next morning, I started the show and opened the phone lines to callers. Sure enough the first caller was from Poughkeepsie New York. Fortunately, they wanted to talk about one of my favorite topics: measuring the arrival of spring via ground temperatures and growing-degree days. We measure growing-degree days with a simple calculation, where you take the high temperature for the day and the low temperature for the day and add them together. You then divide by 2, and, if the number exceeds 50 degrees Fahrenheit, you have amassed some "growing-degree days." This helps us anticipate insect activity or bud break as the earth comes to life. If you have a spring day that struggles to a high temperature of 64 degrees and a low temperature of 39 overnight, then $64 + 39 = 103$, divided by $2 = 51.5$. Hallelujah, we accumulated 1.5 growing-degree days today which wouldn't excite my friends in Kissimmee, Florida but is cause for jubilation here on the frozen "pundra."

Even more fun than math is the experience of jabbing a soil thermometer in the ground on an early spring day. White snow is the great equalizer. Everyone's yard looks the same under those conditions. When the snow retreats, it is a rite of passage to find a patch of green and press a soil thermometer into the wakening earth to measure its temperature at the two-inch level. It's a daily occurrence, and we live for the needle to move. Once soil temperatures reach 60 degrees everything breaks loose. It's a garden party. I've been caught pulling over and probing someone's yard on an early spring day. It's difficult to explain but all part of being grounded my friend.

I REMEMBER THEM WALKING IN THE ENTRANCE ONE DAY ON a busy Saturday afternoon … a couple looking lost. She was dressed very nicely, wearing something you would not wear to dig holes for begonias. He had a sheepish look on his face and didn't say a word. It looked like he had just come from the golf course wearing a light-yellow polo shirt and white shorts with dress shoes. It was obvious she was in charge, and he had been dragged to a place he didn't want to be.

They approached me and with a wave from her hand her first words were, "show us the plants that don't require any care. We want something that will look nice and just want to get this over with." He nodded in approval behind her because it seemed that was the thing to do. I brought them to the silk flower section and told them to have fun.

It's fun to be a superhero and help people pick out the exciting, perfect new plant for their yard or home. My kryptonite however is the difficulty in focusing on just one plant. Every plant is my "favorite plant" and I'm like a kid in a candy store. I'm a firm believer that we as plant experts need to spend more time telling people "why" they should garden as opposed to "how" they should garden.

Picking out plants is a metaphor on finding direction in our life. When we don't know what to do or what direction to take, that's when our real odyssey begins. If your mind is not befuddled, then you're not truly engaged. If you're comfortable, then your adventure has not begun. How many times have I heard the phrase, "I have a question for you." Let's dig in. Through the years I have always considered in my work that the greatest compliment I could receive is from someone who says they listen to the show but don't have a garden. Lessons are to be learned in the process. The plot thickens.

Plants are fashionable and arguably an art form that are appreciated for their beauty. Poor Wilma Plantsgrow doesn't realize that, in essence, she is a present moment case study in Gestalt psychology when she tries to pick out plants for her landscape or pot. Her world and our world is the mind's ability to make sense of that world, based on a fragmented knowledge of its parts. If the world is the sum of its parts … then why garden? With a little bit of knowledge, we take ownership of the arrangement of those parts in a way that makes our little corner of the world a better place.

Overcoming the fear of failure is the first step or hurdle to clear in the mind of Wilma Plantsgrow. I attended a garden conference where the following statistic from a study was shared. For three percent of people who buy a plant, it dies the same day. Really? That's like a kid that tires of a toy by Christmas night and is playing with the box. How do these people get dressed in the morning and go out in public managing to keep themselves safe all day? Sometimes people walk up to me with a plant in their hands, extend it towards me and ask, "is this plant going to live?" My initial thought is *if you put it back it will*. They shake my hand vigorously and say they watch me on TV or listen to the radio show. The next words out of their mouth are, "I have a brown thumb, everything I touch dies." My initial thought is *that is a piece of information I would have liked before you shook my hand*.

There is a Chinese proverb that says, "The best time to plant a tree is 20 years ago. The second-best time is now." Don't use the excuse that you don't have a green thumb. That proverb could apply to any good you can do in your life.

Most people approach a gardener with self-depreciating humor or a jab at themselves. They throw themselves at the mercy of a horticulturist to save them from their "I can't get anything to grow" reality. There are studies that suggest that by the year 2030 as many as 375 million jobs globally will be lost to automation. These studies are often quick to point out, however, that jobs involving managing people, social interactions or applying expertise will be spared. In the next breath these studies mention "gardeners" and "plumbers" will be spared from automation. Makes me feel good that Wilma Plantsgrow is going to need the expertise of a gardener for years to come and not a robot.

I play word association with groups I speak to, and when the word "garden" pops up people associate the word "work" as an impulse. I prefer exercise or fresh air or natural experience in association. It's only natural, when you think of it, because we live in an environment of contradictory tendencies and opinions. It's what makes the world go around, or, as they say, provides the spice to life. People relate better to the word "landscaping" than they do to "gardening." This might be true. In an impromptu survey I Googled the word "garden," and the top keywords in that search were "where is the Garden of Eden" and "what time does the Olive Garden close?"

Now I understand that admitting failure is a great teacher. If you haven't killed any plants, you're not trying hard enough. There are no gardening mistakes, just experiments. You must be able to laugh at yourself and your failures. Success is not a good teacher ... sometimes being hungry makes you sharper and more fully alive. Losing teaches you a lot ... winning, not so much. Hunger makes you less risk-averse. In the Great Depression, during the years of 1929 to 1933, life expectancy actually *increased* by as much as six years according to researchers. Life expectancy rose from 57.1 years in 1929 to 63.3 in 1933. Imposed lifestyle changes and a hunger for something better seemed to have a counter intuitive effect on what we thought would be the case. Don't get me wrong. Hunger worldwide is a serious problem, and I am not referring to malnutrition. I am talking about a mental hunger to change, to grow, to improve. As opposed to a sated condition, an edge can have a positive effect on outcome almost every time.

> Sometimes you have to jump and figure out the landing on the way.

I remind myself that numerous mistakes need to be made, because the Hindu saying is right, "When the student is ready the teacher arrives." Time and again I have learned in life that failure is not aiming too high and missing, it is aiming too low and hitting it.

If you are one of those people who must be right, there's a

good chance you will be risk-averse. From time to time I have someone who asks me a question, but I can see in their eyes they already know the answer. They want to tell me what the answer is but preface their end goal with a question. I've learned that I'm not there to be right … I'm there to get to the right answer. Sometimes that process is humbling. Be humble or be humbled.

In the gardening world, seeds and bulbs are notorious for being purchased and never planted. A large percentage of well-meaning individuals proceed to leave them in the trunk of their car or on a shelf in the garage. Six to 12 months later they call me to ask if they can still plant them. I am one for winter planting of bulbs like tulips or daffodils. From time to time, in December, you can get flower bulbs on clearance that are just like me, 75% off. I can't resist the deal with my Dutch roots, so I buy a bag full. It's like when a fast food restaurant offers two sandwiches for the price of one. I couldn't force down two of them, but I would be crazy to pass up on this deal.

Winter bulb planting with clearance bulbs was an annual tradition for me in my garden. My neighbors watch with curiosity as I take a show shovel and begin clearing snow from the landscape. I then use a pickaxe to break through the upper crust in the ground. With the winter solstice occurring in the month of December there are few daylight hours. One evening my neighbor came home from work, and there I was in the yard with a flashlight, snow shovel, garden shovel, wheelbarrow and a pick axe. He paused as he stepped out of the car, and straightened himself warily, observing from a distance. We stood in embarrassed silence, looking at each other with only the sound of our breathing, and the mist of our breath filling the air. I was filthy, covered in dirt and snow, holding a pickaxe. Because of the cold temperatures, I had one of those black pull-over balaclava masks on that people wear when robbing a bank or a liquor store. I had dropped my flashlight, and it cast long, dark, elongated shadows across the snow. Instead of trying to explain, I decided to proceed with the pick axe as he quietly and briskly stepped into his home. We never discussed the moment, and he moved a short time after the event. To this day I'm sure he thinks I'm a serial killer.

Here is what I love about gardening. Unlike the movies, the content is appropriate for people of all ages, ethnicity, sex, gender and culture. Everything is rated G for gardening. So, I love the concept that we are not millennials, Gen X or Y or baby "bloomers," but rather we are one in the garden. We are in the garden of "needin." All of us. We need each other. We need understanding. We need fresh air to breathe. I have learned unconditional love from the garden. We are the Perennials. I have friends who remind me that age is just a number. For that matter, any category is just that, a category where evaluation and grading supersedes just living. Together we band to make the world a better place. I love that spirit and it lives in the grounded spirit of gardening. I think of all my friends, brought together by the common cause of making the world a more beautiful place, and I marvel at the diversity. In the garden, where they came from or where they are going, doesn't matter as much as if the sun is going to shine today. And we all share the same sun. My heart has broken throughout the years, because a person fits in a certain category and is evaluated or criticized because of it. If all are created in the image of God, then the perfect place to celebrate that is in the garden right? So, no matter how you are labeled, you are welcome here.

Seven percent of all American adults believe that chocolate milk comes from brown cows. If that doesn't seem like much, remember seven percent of all adults is millions of people. These are the same people who are making watering decisions for their plants. Life-and-death decisions for their ficus. Many people want me to give them a schedule to follow when watering their plants. They insist that if I share the secret with them they can handle it. They want me to tell them to water every Tuesday at 9 PM. Or, to water twice a week before noon except when there is a full moon. They want a process they can schedule on the calendar or in their smartphone.

Plants, like people, have differing needs. The environment also changes around them, depending on the day, week, season of the year. With houseplants water is the number one killer ... usually too much water. Most people are too nice and kill their plant with kindness. They pour on the water and rot the roots. If day-length shortens or there isn't a sunny window, the plant simply does not transpire like people perspire and need the water. If you were forced to drink a two-gallon jug of water when not thirsty you would feel a little rotten too.

I have a friend in south Florida who subjects houseplants to indoor duress to check their limits. I love visiting southern Florida every year to hang out with the plants. I don't like snow, so some foliage and sunshine in January rejuvenates my spirits. I call it lifestyles of the "Rick and ramous." For those of you questioning my sanity, ramous is an adjective meaning relating to or resembling branches. I like to go out on a limb, because that's where the fruit is. She did an experiment with a *Dracaena marginata*, putting it on a water fast until some stress began to show on the foliage. It took 53 days before some yellowing and tip damage began appearing on some of the foliage. Granted, a *Dracaena marginata* is a tough guy in the plant world, but it points out that we tend to overwater our indoor plants. Kill them with kindness as in too much water. Watering needs vary both on the species of plant as well as the environment it dwells in. Regardless, if the top of the plant is not transpiring and calling for water from the roots, the roots will rot and choke from a lack of oxygen.

Learn how to water based on the weight of the pot. Or, dig a hole, fill with water and stand back to see what happens. Plants have varying water needs. A succulent needs to dry out between watering. They wait patiently with their fleshy foliage for the next flood. A palm grows best if the soil is not constantly made wet in the upper profile of the pot. Rain storms in the tropics are often a deluge when they come. With a palm, allow the soil to dry between watering, and then when you water, pour it on so it is coming out of the drainage holes in the bottom. Ficus will often drop 50 percent of their leaves and pout when summer is over. They throw a temper tantrum when the day-length shortens, and the warm, sunny weather gives way to fall and winter. Let them get it out of their system with a petulant position in the corner of the room. Adding water is like adding fuel to the fire and will speed up its sullen situation.

A plant's roots need oxygen as much as they need moisture. Root depth is a consideration as some plants are buried alive as opposed to planted.

Winter sun, for outdoor landscape plants, is a factor in northern and Midwest gardens. The sun positions itself low in the south horizon, and, combined with wind, can desiccate plants with south or west exposures. With broadleaf evergreens, like rhododendrons, the ground freezes, and the plant is unable to get a drink when exposed to the sun and the wind. The age-old question of whether or not it's a sun or shade plant is often less important than "what side of a structure it's planted on."

I find a lot of people struggle with the issue of north, south, west and east orientation. When selecting plant material, it is good to first determine what direction that side of the structure faces. Here is how the conversation usually goes.

"I want some plants for the front of my house and need some suggestions."

"What direction does the front of your home face?"

Silence. Thought.

"The street."

"No, the direction it faces,"

"My neighbor."

"No, is it north, south, east or west?"

27

More silence. More thought.

"Which way is north?"

These are people who would follow their GPS off the end of a pier. Finding their car in a parking lot is a big achievement, and they often get lost trying to find their table in a restaurant after using the restroom. When they ask for directions, telling them to head east is no help whatsoever. They need a big mall map with a large YOU ARE HERE location star.

Twenty-five percent or more of the questions asked in the 25 years of answering questions are lawn related. We love our green, cool-season lawns here in the Midwest and Northeast. The "lawn" and short of my conclusions after helping answer the thousands and thousands of questions for years is what? When I am asked to speak to a group, they generally want me to leave time at the end to answer questions, which means, in essence, my presentation has caused even more questions in their mind. I'm okay with that, because I understand that when I speak to a group, they will forget most of what I said in short order. What they will remember is how I made them feel. That's what they walk away with. The success of the speech hinges on how I made them feel. The same can be said with their gardening question or project. Two friends have helped me sort this out. They remind me that people want me to make them a hero, and, secondly, I need to start any advice with the end in mind. That has caused me in my older age to take the following approach:

I now ask "Why" more than I ask "How."

Asking why gives what we do purpose.

When people have purpose, they become everyday heroes.

Sometimes people just need to know *why* they should garden and not always *how* they should garden.

When someone has an embarrassing appearance due to cup-cake frosting on their lip, something hanging out of their nose or their zipper is down, do you tell them? This topic came up on the show one week when I had eaten a donut and had some frosting on my chin. No one told me. Who knows how long I interacted publicly in that condition, until, finally, at some point I visited the restroom and discovered it on my own. I asked the question "would you tell that person?"

Linda contacted us, and told me a person should be told if the situation can be easily remedied. For example, you have some salad lettuce on your teeth or your lunch is stuck to your face. A person should not be told if the issue cannot be remedied. If you have a mustard stain on your shirt or you have a very bad haircut. The point is to help and never to embarrass, which would not be proper etiquette.

So, what if your neighbor's yard looks awful, and their latest horticultural additions are hideous? Do you say something and offer your opinion, or do you just let it slide? If their yard ornaments are offensive, their lawn care is lacking and their anemones atrocious, do you say something? If they could easily rearrange, with a little encouragement, do you broach the subject? Or, do you let them go on their way with their plants down?

I'll post pictures of pretty flowers I spot on the roadside when I'm running. A friend posted this in response:

> *"Do flowers just sprout where you pass?*
> *Kind of like a Disney princess or something?*
> *I think they bloom just to impress you, but if*
> *birds start landing on your outstretched hand*
> *and woodland animals start singing to you,*
> *better get to a doctor!"*

My response was:

> *"I think I'm suffering from Disney spells."*

So, when Wilma Plantsgrow asks me "how do I become a green thumb? How can I be a gardener? I have my top-ten list ready for her.

1. First and foremost, kill some plants. If you haven't killed any plants you're not trying hard enough. Remember the goal is not to never fail. The goal is to fail better each time in the quest to better skills and results. You're going to get better in life if you can fail in the following manner:

   • Fail at first

   • Fail repeatedly

   • Fail judiciously (Don't buy the farm. In other words, if you fail but don't end up in jail or dead or hurt anyone else, you can consider it a success.)

   • Own your failure. Then move on, don't carry it with you.

2. Garden in October and November when few others are. Get some bargains and try them out. It may become a beautiful focal point on your plot. Even a blind squirrel finds a nut now and then. Use weed controls in October and November. The most effective time of the year.

3. Buy some plants that work three to four seasons for you. They work even if you're not working.

4. Work some low-maintenance plants into your landscape. Accept compliments on their robust growth, while all the while knowing you're not doing a thing.

5. A landscape or an epic garden, like most great changes, were preceded by chaos. Pace yourself. Rome wasn't built in a day.

6. The artist Vincent Van Gogh, who I prefer to call Vincent Van "Grow" because that guy obviously had a green thumb, said "Normality is a paved road ... It's comfortable to walk on but no flowers grow on it." Be willing to experiment. Don't be a cookie-cutter

neighbor.

7. Look down not up. Start with the planting area and soil. Don't put bald tires on a $50,000 car.

8. Get a houseplant before you get a pet. Practice. Learn to nurture. Understand that most houseplants are killed with kindness. Water (too much) is the number-one killer of houseplants. Learning to care for houseplants teaches us a lot about ourselves.

9. Remember gardening is a four-season activity. Spring is just a holiday on the calendar. For example, a lot of important pruning takes place in winter. Don't be afraid to prune and use the rule of green thumb, prune after bloom. Whether a rhododendron, lilac, forsythia, azalea or althea use the standard "prune after bloom" approach. If you prune deciduous plants in winter, you get a good look at a plant's shape, disease and insect activity. (It's better than watching re-runs on TV.) Dress warmly and remember, landscaping is not just an activity for the month of May.

10. Think exposure. You wouldn't leave the house in the morning undressed. Think of sun and shade as a four-season thing. A rhododendron can thrive in the summer sun but will wilt like an actor who forgot his lines in winter sun.

His lawn remained uncut
He had a copious lime ball glut
Mower blades would sectile
and spew the projectile
Blame the insidious green walnut

Chapter Three

My Mom told me there would be daisies like this

I REMEMBER THE CLOTHESLINE POLE AND the lines of clothes fluttering in the breeze of my boyhood backyard. At Grandma's house, the flutter of the unmentionables, hanging on the line without embarrassment, was akin to flying a flag of surrender, because we've all given up just a little on life. I mean, let's face it, if you are willing to pin your intimates on a rope for all to see, you're making a subconscious statement. Or maybe you're trying to show off? In today's day and age, if you hang the undies on the line for all to see, your skivvies may find their way into a social media post. We can be thankful for dryers restoring modesty to the landscape and backyard. I can see the environmental benefits of hanging out your laundry, but a sunny secluded spot would be in order. Our neighbors Bud and Ruth wouldn't even try to disguise their laundry; their clothesline pole was planted in the front yard. It was socially accepted behavior in the '60s. It remained there until my sister backed over it with the car. At least my parents strategically placed our clotheslines in the backyard behind the garage. This turned out to be a flawed strategy, due to a large and rangy mulberry tree growing along the property fence.

Mulberries have a long and storied history and are found throughout the world. In some parts of the world the tree, namely the white mulberry tree, *Morus alba*, is used for silk production. Silk moths lay their eggs on the foliage, and the hatched larvae feed continuously on the foliage, molting as they go. The silkworm, native to Asia, creates a cocoon, and they enclose themselves within raw silk produced by salivary glands. The pupa is boiled to death, dipped in hot water, so the cocoon can be unraveled for silk thread. In ancient times, China was renowned for its production of silk, and, like the clothesline pole, has become less popular, having gone the way of synthetic fabrics.

The mulberry tree has been prized by cultures for centuries. Even Shakespeare in *A Midsummer Night's Dream* has the lovers meeting beneath the cover of a mulberry tree. Mulberry trees were brought to the United States in the 1700s with hopes of development of silk-industry production. In the late 1700s, commercial nurseries made them available with two very famous customers, when George Washington (Mount Vernon) and Thomas Jefferson (Monticello) planted them on their properties. Prized worldwide as a food for both humans and animals, it has been widely used for medicinal purposes also. Therein lies the problem. As the tree escapes cultivation, especially the black mulberry, copious amounts of pollen can be produced, resulting in, for some, respiratory issues and sneezing. In addition the tree is very fast growing and weak-limbed, making it a less than desirable landscape specimen. The mulberry tree, like cilantro, is a polarizing topic on the radio show, so, when I bad mouth it, I can plan to receive rebuttals from its fans.

Birds love mulberries and are expert propagators of the species. They seemingly evacuate the contents of a mulberry as quickly as they eat it. I remember watching this phenomenon as a boy when they would dive bomb my Mom's laundry hanging on the line. They seemed to find avian enjoyment, using the sundry, clean linens as a target, an invitation to strike with amazing accuracy. Of course, the bed sheets were always the coup de grace target of their shenanigans, making an artful colorful splash on their surface. I was amazed at their accuracy, while seemingly drunk from the bounty of berries they would consume.

Memories of the clothesline and the clothesline pole as a kid seemed to resonate with our radio show listeners. My primary memory was how I used the pole to do chin ups as a kid. When first learning how to ride a bike, I crashed into the clothesline pole and did a lip stand on a post. But, mostly, it is the nostalgic memory of Mom hanging laundry on the line on a breezy, warm summer day. We would borrow the clothespins to clip baseball cards on the frame of our bikes, so the spokes would make a racket as we rode. It sounded like something between a motor and a roulette wheel until the card wore out. I always wanted a Mattel Vrooom, a battery-operated motor sound you attached to your bike, requiring four, D-cell batteries. It never appeared under the Christmas tree, so I settled for stale gum and baseball cards every summer.

As an adult, I think of the old clothesline when I watch people who are hung out to dry. Some people tear down others in order to build themselves up in their own minds. There is a difference between healthy competition and leaving someone out to dry, pulling the rug out from under them, dropping them a few pegs. When you compete by raising the level of your performance everyone benefits. Rising tides float all ships. When you tear someone down or undermine them, it lowers the level for everyone. Why some people think they are elevated by damaging someone else is mystifying to me.

> When you compete by raising the level of your performance everyone benefits. Rising tides float all ships.

I guess that's where grace comes in. I have learned that grace is a beautiful thing, and extended grace is like water from the garden hose; it flows to the lowest level.

Some plants undermine others by, in essence, inhibiting the growth of their nearest competition. Root competition, shading, competition for moisture and nutrients, I consider all fair game in natural survival of the fittest. We plant our plants, taking these things into consideration as our "babies," giving them the best nurture we can offer. There is always a hunger for that next, best fertilizer or amendment we can use to make both us and our plants look brilliant. We look for those advantages, so our plants are rooted in success and given all the opportunity we can offer. We proudly watch them grow and accept compliments on their robust growth. We are awarded with the recognition of being a green thumb by our peers, or at least by a spouse and neighbors.

Other plants undermine through a more insidious means. They sabotage other's growth and diminish other's standing in the landscape. I have learned in life that, just as I don't like those people, I generally don't like these plants. The plants are considered allelopathic plants, where subvert is the name of the game they play. Plant allelopathy is when a plant uses chemical warfare to suppress others and then take advantage of them. This is not team-play in the neighborhood. Allelopathy comes from a couple of Greek words, *allelon* meaning each other and *pathos* meaning to suffer. An example would be the misnamed invasive species tree of heaven, *Ailanthus altissima*. You'll find this tree growing along roadsides, ditches and any place else you don't want it. I spot it when out running and it is often mistaken for sumac. The common name 'stinking sumac' is more appropriate for this plant. If you take the time to get to know it, the flowers and foliage smell like rancid, bad peanut butter. I have often wondered if peanut butter goes bad as I drag a jar from the back of the cupboard. I guess the presence of a "best if used" by expiration date reminds us it is not guaranteed immortality. I mean everyone knows if you stir it, then it's good to go. If you can imagine the worst-smelling peanut butter in your mind, this tree smells like it. I got sick once as a kid on a peanut butter and jelly sandwich. I steer clear of tree of heaven, because the aroma reminds me of that episode.

Even some beloved plants that conjure up images of health and wholesomeness can succumb to allelopathic behavior. Broccoli residue, as an example, can inhibit related plants like cabbage or cauliflower to grow in an area previously planted with this revered green. This dark side of broccoli may come as no surprise to those kids, as well as adults, who have to explain why they didn't clean their plate. Now, of course, you can ban broccoli when you're President of the United States as George HW Bush famously did in 1990. George Bush told reporters his mom made him eat broccoli as a kid, but now he as President didn't have to eat it anymore. He banned broccoli from Air Force One, and I think the vegetable has sought its revenge ever since.

The best intentions of the misguided souls who brought garlic mustard to our borders, did not foresee its allopathic tendencies to crowd out desirable plants, lawn ornaments and small pets in its path. Even good people and plants can go bad.

Competition is a good thing, but, with plants, like people, to undercut, damage and diminish another for their own advancement, is well … just nasty. It's toxic. When it comes to this behavior, the allelopathy role model, centerpiece and poster child has got to be the black walnut tree, *Juglans nigra*. Its notorious and ubiquitous presence has become, as the French would say, a cause celebre on the radio show. I stir dissension when I say I do not like black walnut trees, and, like George HW Bush and his aversion to broccoli, I'm willing to go public with my disapproval.

Black walnuts produce a chemical called juglone, that is present in all parts of the tree, but especially in the nut hulls and in the roots. Symptoms of black walnut toxicity include leaf yellowing, wilting, stunted growth and eventual death. Many plants are sensitive to juglone, and sensitivity training is not about to help. They create an adverse environment that is toxic to many plants in relative proximity to their surroundings … a quiet, insidious, toxic behavior that slowly but surely erodes life around you. Sound familiar in some workplace environments? You can probably name the "black walnut" personality hanging out at the coffee pot in your office.

The black walnut was part of Native American diets, but most of the walnuts we eat today are english walnuts. They are milder and have thinner, easier-to-crack shells. Black walnuts have an unmistakable earthy flavor ,which is a nice way of saying "please don't ruin my fudge by putting them in there."

If its toxic behavior towards others isn't enough to make you dislike them, their fruits inimical behavior will turn you against them. Black walnuts have a lime-green husk, and they look somewhat innocent like big, round limes hanging in the tree. That's where the good feelings end. The nuts will stain the dickens out of everything, from the dents in your car to the sidewalk and driveway, when they begin their bombing sortie on your property. I wouldn't even park a rental car under a black walnut tree. The lime balls nestle in your lawn and will trash your mower, becoming projectiles that might break a kneecap or a window in the process. Step on them before they become brownish, slimy goo and you'll twist an ankle. The husk contains the chemical called juglone, but even worse than its toxic behavior or ability to stain the daylights out of anything, is the aroma. With their bitter, pungent, distinctive aroma, these odd-smelling lime-nut balls smell like blue cheese gone awry. They smell like a musty room with bad cheese mixed with a twist of citrus, or a gym bag with used workout clothes, old running shoes and spilled energy drink left zipped in the bag for a month, sitting in the back seat of a hot car.

As kids, in September and October, we would use shovel handles and hit them like baseballs in the field for hours on end. Our nostrils had not yet developed the sensitivity needed to abstain from this behavior. We would fantasize, stepping up to the plate and getting a hold of one, for a game-winning, walk-off home run in the World Series. After batting practice, you could pick the lime-green shrapnel and bits from your arms and face. I'm sure Mom was disgusted by the aroma of our clothes that ended up in the washing machine. Maybe she threw them out, I don't remember. We had fun and the lime orbs also made for epic fruit fights until someone took one to the forehead.

Black walnuts shed their leaves, nuts and copious amounts of twigs and branches in September and October. I owned a home with multiple trees on someone else's plot adjoining my property. I schemed in my mind how I could eliminate the trees without anyone knowing, even though they were just over the property line. I would sit in my chair in the dark, looking out at those trees at night, conspiring a strategy to deploy. Rocking back and forth, staring off into space … I had crossed over to the dark side. Believe me, it took every ounce of diplomacy I could render to keep from starting a ground war. All the neighbors loved the trees for the shade they produced. I was a lonely rebel with a cause left out standing in the field. The neighborhood watch group was watching my every move. The siege dragged on from weeks to months to years. They saw shade and beauty … I saw firewood. The squirrels would sit in the trees chattering and laughing at me.

I could deal with the neighbors but not the squirrels. Squirrels go crazy in September and October. If not crazy already with their antics, dodging traffic in the middle of the road and chewing on electrical wires and propane gas grill hoses all summer long, in fall they go nuts on this lime-green fruit. Squirrels have sharp teeth and are persistent. When it comes to black walnuts, anything I have to drive over to hull with my car in the driveway before I can eat it, is too much work, and I certainly don't have the patience. However, squirrels have nothing else to do, aside from raiding the bird feeders, so they will hull nuts all day long. Squirrels are excellent propagators of black walnut trees. I've observed them rubbing the nuts against their head to mark their scent on their find. They then plant them everywhere, only to lose their placement when hungry in winter. The end result is a bumper crop of new walnut trees the following year, growing in your garden, your lawn and out of the foundation of your house.

As you can tell, I use my view of black walnut trees as an analogy to toxic relationships in my life. Having to deal with certain people I dislike, I simply envision them as a black walnut tree. It truly is quite helpful. You should try it. When they start to seep their juglone in my space, I picture them as a big walnut tree. They can bark, go out on a limb and bomb me with the fruits of their attitude. I move beyond the shade they cast. Negativity, suppression, subversion and sabotage are all toxic behaviors, and when you hang around you become a product of their environment. Next time someone toxic tries to shade your day, picture them as a big, old, gnarly, messy tree and "leaf" the area.

> Next time someone toxic tries to shade your day, picture them as a big, old, gnarly, messy tree and "leaf" the area.

My mom taught me that principle from a young age. She was always energetic and either whistling or singing as she worked. When issues arose, she would take the "it was meant to be and everything is going to work out" attitude. When you deal with toxic conditions in your life, remember it's nothing a mom can't fix. Studies have shown that a conversation with a mom, when stressed, can reduce key stress hormones, and, more importantly, increase oxytocin levels, a feel-good chemical associated with empathy and love. Cortisol levels quickly drop when a mom is around to listen or talk to. So, when stressed, due to toxic surroundings in your life, first envision the source as an old, gnarly walnut tree, then find a mom to talk to. You'll feel much better. Take it from me; my mom told me there would be "daisies" like this.

His hydrangea would simply not bloom
and caused him to fret and fume
Despite fertilizer
Being all the wiser
The weather had sealed his doom

Chapter Four

That is so Bosky

bosky: *adjective* \\'bäs-kē\
having abundant trees or shrubs

FIRST SAW HIM AT A DISTANCE LEANING against a stack of fertilizer bags. He would appear and then disappear without notice and had the grin of the Cheshire cat in Alice of Wonderland, the kind of grin where you knew conversation was going to be interesting but never really serious. He would show up bearing gifts, one time a pickup load of ear corn for me, and the next a case of his homemade wine. This wine packed a punch. The first glass was for health benefits. The next was to demonstrate your dancing prowess whether invited or not. Paul was like me, an entre-manure and a man on his own unique mission. Paul's specialty was ice cream. We had that in common as business owners, because ice cream is like flowers, and who doesn't like them?

Paul approached me with a business proposition. It didn't take much to convince me. He wanted to name an ice cream sundae after me at his ice cream shop, putting me in competition with others who had done the same. A local meteorologist, a politician and a sportscaster had their sundaes posted on the menu board, and it only made sense for me to enter the arena. A culinary cream creation was born on the spot. The mulch pile sundae was born. If you opted out of a sundae, and a flurry was your speed, they would use their magic weed whipper and give you a compost pile flurry.

One of the elements of the mulch pile sundae was a habit I had since I was a little kid. I always added Rice Krispies to my ice cream. I wanted a little snap, crackle and pop to my creations and this was no different. Rice Krispies have been around since 1928, and, as far as I was concerned, their breakfast contribution was secondary. Rice Krispies were meant for greater things … the ice cream sundae. It was born out of a visit to Battle Creek and the Kellogg's factory as a little kid. At the end of the tour, there were little cups of ice cream, and you could add cereal to your dessert. The combination of snap, crackle and pop to the creamy goodness of ice cream was a delectation to my taste buds and senses. In my younger days, when sugar was a staple in my diet and considered a major food group, it was Cocoa Crisps for breakfast and Rice Krispies on ice cream before bed, slathered in chocolate syrup. Who knew that inflated dried puffed rice could taste so good to a growing boy? Those days are gone and my consumption habits have changed, but they live on in the form of the mulch pile sundae thanks to Paul.

Our consumption habits continue to change, and not just in our daily lives or our diets. Instead of consumption of stuff, we are more in tune with experience, and the same applies to our home landscapes. We want plants that provide more than just one season of interest, plants with a purpose that are durable and non-invasive.

The statistics from national garden surveys continue to point out that a greater percentage of households are participating in lawn and garden activities. I find the word "activities" rather humorous and difficult to define. I'm sure the people who participate in these surveys define lawn and garden activities differently. Weeding, planting and pruning may be for one while lounging in a lawn chair with a cold adult beverage might be another's form of lawn and garden activity. Are you strolling behind a lawn mower or dividing your perennials and splitting your plants? Are you digging holes to plant hydrangeas or is grilling a ribeye steak on the deck a lawn and garden activity? I'm just glad that the great outdoors is now considered an extension of the home, an outdoor living room for activity and experience.

I like feeling bosky. No not that unshaven feeling where you have the scruffy beard of a sailor at sea, but, rather, that feeling when you are aware of the changing seasons because of deciduous shrubs and trees. We love changing seasons. It's like a "do over" or a renewal when the lilacs are in bloom or the viburnums turn color in fall. Landscape shrubs and trees are wonderful at marking the seasons of our lives. You might not realize it, but deep down you like feeling bosky too. Plants help people celebrate the changing of the seasons.

> For some they face it, for others they embrace it.

In the Midwest and the North, we have very distinct and identifiable changes to all four seasons every year. For some they face it, for others they embrace it. Deciduous plants embrace the change of seasons, moving from dormancy, to spring blooming, to summer growth, to fall color and back to dormancy again. Fortunately, dormancy is a reversible condition. Deciduous shrubs and trees both face it and embrace it, and we are the beneficiaries of the gift of seasonal change that they exhibit.

People who pay it forward respond to a person's kindness to oneself by being kind to someone else. That is how we pay it forward. In essence, plants are natural givers too; they pay it forward.

Plants, if given the benefit of a good place to root, a few nutrients, sunshine and water, they in turn provide oxygen, mood-enhancing support, color, food, seasonal change, shade, wind block, protection, aroma, medication, flavor, flowers, scent, air-cooling, nature therapy, privacy, wildlife shelter, canopy, ground cover, erosion control, pollinator nectar, building materials and much more.

Trees, shrubs, flowers and grasses, truly know the meaning of paying it forward. And flowering shrubs from hydrangeas to rhododendrons are seeing a resurgence in interest after years of herbaceous perennial dominance in design, because of their sequence of bloom in the garden. Flowering shrubs can be low-effort and big-reward, hard-working plants for the homeowner landscape. When a rhododendron blooms it can be rewarding, your hydrangeas can make you a hero and your lilacs can be uplifting. A vibrant viburnum in bloom doesn't ask much in the way of effort from its owner, but every year delivers a knockout show. Landscape own-root shrub roses in a sunny spot make the world a brighter place in summer and fall, and give far more than they receive with season-long color.

Woody deciduous shrubs can be damaged in winter by ice and snow, exposure or deer, vole and rabbit browsing. For that reason, adding some herbaceous perennials to the mix that *act like* shrubs in summer, but die back to the ground in winter, can be a valuable addition to your landscape. These perennials quickly grow large in summer, making a visual shrub-like impact, adding gravitas to the scene. Because they die to the ground in winter, they are not subject to the animal-browsing or winter-exposure elements. My favorites are:

- Ornamental grasses
- *Baptisia australis* known as false indigo
- Perennial hibiscus
- *Rudbeckia maxima*
- *Helianthus maximiliani*
- *Heliopsis helianthoides* (false sunflower)

- *Persicaria amplexicaulis* (mountain fleece)
- Large hostas like 'Frances Williams' or 'Krossa Regal'
- *Ligularia 'the rocket'*
- Cimicifuga *Actaea racemosa*
- Canna
- Goatsbeard *Aruncus dioicus*
- Joe Pye weed *Eutrochium purpureum*
- Russian sage *Perovskia atriplicifolia*
- *Acanthus spinosus* known as Bear's Breeches

My favorite flowering annuals for that summer "shrub-like" size would be *Cleome*, *Amaranthus*, castor bean, *Nicotiana* and *Verbena bonariensis*.

When it comes to pruning woody, flowering shrubs or flowering trees, there are a couple of rules of green thumb that will serve you well in your endeavors. First, do not be afraid to prune. Most flowering shrubs and trees welcome a good pruning, and the increased light and air penetration will make them all the better for it. Second, if you question when to prune, apply the general rule that "you should prune a plant right after it is done flowering." Flowering plants tend to work over the course of the year to produce new buds for blooming, so if you prune it after it blooms in its normal cycle, it will reward you with a fuller-blooming plant the following year.

Sometimes rejuvenation is in order for our woody, flowering shrubs and trees. I took a long hedge row of underperforming lilacs at the farmhouse home of one of my viewers on TV, and ran a chainsaw a couple feet above the ground the length of the row. Fortunately, she trusted me and was a willing accomplice, because the lilacs had become thin and underperformed her expectations. What did we have to lose? Sure enough, by the time the next year rolled around, fuller, healthier and robust, blooming plants had replaced the stretched and tired plants that had frustrated her. I was a hero. It was again a metaphor on life. Fat, sassy and successful was no way to go through life, because setbacks and adjustments make life all the more interesting and rewarding. What fun is life without peaking the crest of the roller coaster, wondering what awaits around the next bend? Sometimes you need to be kicked in the rear and watch how you respond. I've learned to respect and appreciate the pruning experiences I've had in my life as both learning and rejuvenation events. The trick is to live through them to teach others and be a better person because of it. An established pruned plant that has weathered a number of seasons is a wise and learned plant. That's my theory and I'm "sticking" with it.

There are so many great deciduous shrubs at our disposal today for our home landscapes. I would have to say, if forced to pick a favorite, it would be viburnum, with its leathery, textural foliage on sizeable plants, making it a long-term statement in the landscape. Viburnum also has sensational and abundant blooms, which on some varieties are intoxicatingly aromatic, and also boasts berries that look like dramatic ornamentation and are a magnet for birds in the landscape. The fall color is sensational, and their winter hardiness is highly dependable. Another would be juneberry: beautiful, white, spring flowers, followed by berries in June and fall color that is simply spectacular. Juneberries are great for human consumption, that is if you get to them before the birds do.

HYDRANGEA HEAVEN

THEN THERE ARE THE HYDRANGEAS ... THE PLANT PEOPLE love to hate, and the source of more questions on a live-radio, call-in show, rivaling the never-ending mole and pruning questions.

Yogi Berra is quoted as saying, "You can observe a lot by just watching." He makes a good point, as I have seen this play out time and again in the landscape ... not just in the landscape, but in life. I have observed that in life's landscape there are three kinds of people:

Those who make things happen.

Those who watch things happen.

And those who wonder what in the heck just happened.

You've got to get in the game if you want to reap the benefits. Maybe Aristotle said it better:

"There's only one way to avoid criticism: do nothing, say nothing, be nothing."

– Aristotle

So, what to do for fledgling hydrangea enthusiasts anxious for a mid-summer show of color? Carefully nurturing their prized purchase with great anticipation, they await the reward of the large, showy blooms that will amaze their friends and neighbors. When just foliage arrives, and no blooms, despite their care, they frustratingly throw in the "trowel."

Plants, like people, react to their surroundings. There are different types of people. There are different types of hydrangeas. I have watched people blossom and thrive in the right environment, in the right surroundings. This is the "don't try to put a round peg in a square hole" analogy. The same can be said for hydrangeas. A hydrangea, in the right microclimate, will thrive when you get to know their personality. Isn't that why we love hydrangeas in the first place? Plants with personality that can mope, be perky, droopy thirsty, floriferous, temperamental, and, in the case of some, change colors based on the environment. Your surroundings affect your attitude. That's why I like to hang around positive people.

I have learned through the years, that people thrive in a style that is more collaborative than imperious. A collaborative environment that we like to call a micro-climate, also best suits some hydrangeas that bloom on old wood. You scratch my back, and I'll scratch yours. Hydrangeas like sun in summer, protection from wind and sun in winter, a moist but well-drained soil, and some sanctuary from extremes. This puts homeowner and some of the more temperamental hydrangeas in a quid pro quo relationship.

You've got a friend in me

In the realm of microclimates, living on the Lake Michigan shoreline, I have found that hydrangeas and rhododendrons thrive here. Massive plants that bloom seemingly carefree as opposed to further inland. Climate plays a role. Creation of protective microclimates at inland planting sites will greatly improve your success rate. Near the lake, we benefit in winter from heavy snow that provides plants insulation. The lake also provides a warming effect in winter, as the warmer water of the lake moderates winter extremes as opposed to the bone-chilling drops further inland. In spring the now-cold water of the lake makes it cooler than inland, and plants are slower to wake up. This is a good thing, especially in years when inland early unseasonable warm temperatures cause plants to break dormancy, only to be zapped by a zinger frost in April or early May. In summer, temperatures tend to be cooler near the lake, tempering the extremes of summer for both the plants and those who flock to the lake for time on the beach.

There are a number of species of hydrangeas, including some that are native to North America. There are six main types of hydrangea grown in North America, and an understanding of their personalities will bring you success.

Hydrangea macrophylla are also known as mopheads, lace-caps, hortensia, or florist hydrangeas. No wonder this gets confusing to people. I think the best way to recognize them is as "big leaf" hydrangeas derived from the name macro (big) and phylla (leaf). Some big leaf hydrangeas bloom on the previous year's wood (old wood) so avoid pruning them unless you must. Deadheading is fine by snipping off the old blooms at the tip. Today's new varieties bloom on new and old wood, and are known as reblooming or remontant hydrangeas. A remontant plant will produce more than one crop in a season. That said, a protected area from winter extremes is helpful. Issues like deer browsing, shade, pruning or temperature extremes can cause disappointments in blooming performance. I find the north and east sides of the house are a great place to grow a big leaf hydrangea. A big mistake people make is to assume they are a shade-loving plant during the growing season. Not true. You should strive for a minimum of four to six hours of sun per day to get mopheads to bloom to their potential. Avoid the temptation to prune back the leafless "sticks" in your landscape. If you drive stakes in the ground, to surround the plant with burlap in winter or fencing to pack with leaves, don't remove the protection too early in spring. If buds swell in early spring weather, and then we get a hard, late spring freeze in the twenties, you can lose the blooms overwintering from the previous year. You can control the color of some big leaf hydrangeas with a soil adjustment. A soil test is needed to determine soil pH and the availability of aluminum in the soil to turn your mopheads blue. Aluminum sulfate is added to the soil to help turn flower heads blue and hydrated lime to turn them pink. Generally, it is easier to turn some mopheads blue than it is to turn them pink.

Hydrangea paniculata are also known as hardy or pee gee hydrangea, and are hardy to zone 3, making them an easy-to-grow and tough hydrangea.

This species of hydrangea has panicle shaped flowers that look like cones, so we call them panicle hydrangeas. The key is that they bloom on the same year's wood, so pruning timing is less critical. If pruning is needed, prune after blooming, or in early spring, for beautiful blooms in August and September.

Hydrangea arborescens is a smooth operator, known as "smooth" or annabelle hydrangeas. They are native to North America and bloom on new or the same year's wood, making them a reliable and easy-to-grow plant hardy to zone 3.

Hydrangea serrata is a favorite of mine. It's a lacecap hydrangea, also known as "mountain" hydrangea. As opposed to being found in its native Asian element along seaside regions, it is adapted to mountainous ranges. This gives us a tough, bigleaf hydrangea with colorful blooms that tends to be tough and reliable in its blooming habits. The plant does bloom on old wood, so again, try to avoid pruning unless needed. I have had to protect this plant from deer browsing.

Hydrangea petiolaris is a climbing hydrangea. I have it growing on a wooden fence as a "set it and forget it" plant. It is hardy, cold hardy to zone 4.

Hydrangea quercifolia, rivals viburnums as my favorite flowering shrub. Native to North America, this is a plant that gives four full seasons of interest in the landscape in a dramatic way. Also known as oakleaf hydrangea, the plant has oak leaf foliage in spring and beautiful white flowers in early summer. In fall, it has dark burgundy foliage, and, in winter, cinnamon, exfoliating bark that protrudes through the white snow. It blooms on old wood, so try to avoid pruning unless needed, and it appreciates a little winter protection.

10 WAYS TO GET BOSKY

1. Caring for your shrubs and trees is not work, it is "exercise." We live better when we exercise. Twenty-one percent of people would rather stand in line at the department of motor vehicles than garden … how wrong is that?

2. Trees and shrubs want to be planted NOT buried alive, drowned, dehydrated, compacted. Lack of oxygen for the roots can quash your quince, maim your maple, plunder your potentilla and clobber your clematis. Learn how to wet your plants.

3. Focus on the root, not on the fruit. Would you build a

house without a foundation? Would you put bald tires on a new car? Do what's natural and add organic matter. Think woodland floor. Be well grounded.

4. Diversity is the art of thinking independently together. If we plant nothing but the same thing, it is boring and dangerous. Problems spread quickly in monocultures. Embrace diversity.

5. Learn to think like a plant. Making the most of where you're planted and being ready to capitalize on opportunity.

6. Expect three to four seasons of interest from your landscape. Get out there and GARDEN IN THE FALL. Warm soil temperatures, cooler air and natural rainfall, followed by the dormancy sleep of winter, makes fall the ideal time for establishment of trees and shrubs. Don't follow the crowd and be a spring-only gardener.

7. Landscaping in groupings of "threes" and bite-size piece approaches to your work, makes you happier and your presentation naturally appealing.

8. Don't be afraid to prune when needed. Prune evergreens in early summer and deciduous plant material in winter. For blooming trees and shrubs, the general rule of green thumb is to prune right after bloom.

9. External influences put a cognitive load on the brain. Mounded mulch on plants, or pruning into shapes like meatballs and tuna cans, isn't a good idea just because your neighbors are doing it. You can lose your ash together. Do what's right don't follow the crowd.

10. The way of nature is this: something must die so something else may live. Don't be afraid to make some mistakes. Breathe. Enjoy. Repeat. Now you're feeling bosky.

A squirrel in the road would venture
in impulse it would indenture
With speed he would master
Avoiding disaster
A regrettable misadventure

Chapter Five

Not Tonight Deer

THE GATE CREAKED WITH A RUSTY resignation as I pushed it open. It seemed to question my intentions, and, in my mind, reinforced my reluctant entry. Being curious makes you more intelligent … but it also killed the cat. Entry into this world of foliage and flora raised the senses and heightened my young imagination. I proceeded through the gate past the rows of rhubarb, corn and tomato plants. The dill plants danced in the breeze while the radishes poked their heads from the ground to see what was going on. Butterflies perched on foliage, seemingly watching my entrance to their garden playground.

It felt as though I was being watched … I probably was, by the vigilant eye of the old man perched in the annex of the old garage that paralleled the garden area. Mr. Hart patrolled this urban farm plot with a keen and watchful eye for intruders.

I have read, and surmise, that aromas sniffed as a child are embedded in one's memory for life. Imprinted on the mind, they dig up memories when revisited as an adult. The sweet aroma of lilacs or lavender can remind someone of dinner at Grandma's house on Mother's day. As I enter the old man's garage annex, the smell of mothballs burns my eyes. The mixed aroma of unregulated, and now unlawful, garden chemicals for homeowners like Diazinon and Chlordane, dusky wet wood and the unmistakable stench of death turned my stomach. Birds, squirrels, rabbits and chipmunks have visited this plot and never made it out alive.

Mr. Hart was one of my first horticultural influences, and not a good one at that. A mad scientist elderly curmudgeon in blue jean overalls, mixing unregulated chemicals and engineering death traps, he spent his days protecting his garden. I would observe from the safe zone across the street, only occasionally venturing into his world. You could hear his pellet gun go off as he fired at birds from his eagle nest perch directly south of his garden plot. A garden-patch sniper armed with binoculars and a pellet gun, he surveyed the field from holes in the wall of his bunker.

Mr. Hart had a problem. He hated a particular bird, and devised means to eliminate it from his little corner of the earth. The European starling is an invasive non-native bird. Released on the streets of New York City in the 1800s, they, like other non-native invasive species, quickly become a pest to our native natural habitat. Large flocks of starlings are damaging to crops and were marauding invaders to his garden. European starlings allegedly push out native cavity nesters like woodpeckers or bluebirds, and their droppings can spread invasive seeds and transmit disease. They are loud and obnoxious and they crashed the party. Farmers employ all sorts of means, from air guns to netting, to fight back against this non-invited guest of our garden bounty.

I observed the tactics used by Mr. Hart in his garden in his never-ending skirmish, where no means were fowl when it came to taking the field. He had engineered an ingenious basket trap held open with a stick at one end. Tied to the stick was a rope, which twined its way back to his bunker of death and eagles-eye perch. Baiting the trap with goodies for the birds, he would watch with string in hand for visitors to his wire cage. A cardinal or robin or other desirable bird was left to dine and then allowed to fly free. The contemptible starling, however, was about to meet its fate. With a keen eye through his porthole and the steady hand of an assassin, he would jerk the string, pulling the propped stick and dropping the cage, making the starling his prisoner of war. Judge, jury and executioner, he would proceed to step on the trapped bird, killing it and adding it to his collection. Throughout the fall, winter and spring months, this went on as he would collect the birds in the annex, stacking the frozen carcasses like cordwood. When the stench became too much to take and overpowered the smell of chemicals and mothballs, he would collect up these casualties of war and carry them to the garden. Here, he used the birds as a twisted form of fertilizer, carefully pressing the lifeless remains into the soil with his foot row upon row.

On the radio show, I have been criticized for being a tree hugger, appreciative of Bambi, Peter Cottontail and considering chipmunks cute, little, loveable fuzzballs with their cheeks puffed with food. I have wondered if the indelible images, aromas and recollections of time in Mr. Hart's garden, made an enduring impression on my nature today. The undesirable birds and animals that visited his garden of death never stood a chance. Today, I feed a three-legged deer that visits my landscape with carrots, because I can't stand to watch her hobble around. I hold up the carrot and talk to her like some kind of deer whisperer. She follows the back of the pack, so she is always relegated to the leftovers. We meet now and then to share a snack, and she will walk right up to me.

We did win a Michigan Association of Broadcasters award, however, for a segment featuring Kristi's shotgun. We had invited some reindeer to hang out at a promotion around Christmas time, and Kristi was busted for building a deer blind in the parking lot. I wish I had known years ago that doing a "garden show" would require a large share of your time, talking about animals, critters and insects. I would have gone to school to study zoology or entomology. I want to talk flowers, trees and shrubs, but things inevitably veer towards moles, deer, rabbits, birds, chipmunks, bugs and skunks. I even had a couple of our listeners, Ron and Irene, who taught their pet parrot my signature phrase "thank you very mulch."

As opposed to the non-native European starlings, the native blackbird or grackles and red-wing blackbirds, have been subject matter for the radio show from time to time. In my defense, regardless of childhood impressions and scars, don't take out your frustrations on these birds; they are protected by the Migratory Bird Treaty Act. At a time when feathers were highly prized fashion accessories, this was a landmark act in North America for conservation. Birds in the late 1800s and early 1990s were considered an inexhaustible resource. Birds are migratory and move around for local food sources, requiring some protection and birthing the introduction of conservation clubs and the bird-feed industry. Think about those images from the roaring '20s and the feathery plume hats worn by the ladies hanging out at the clubs. My hopes would be that the birds today, like this book, would fly off the shelf.

> Word gets around and birds are adept at tweeting information to their friends.

The red-winged blackbird, *Agelaius phoeniceus,* will take matters into its own hands and attack humans who venture too close to its nest. One of my favorite running paths adjoins a section of marsh area, a favorite nesting habitat for red-wing blackbirds. On my runs along this path, the birds will attack from the back and give me a poke to the back of the head. I don't know if it's their beak or feet, but it startles me enough that I don't stick around to find out. They loudly contest my presence as I approach, and they fire a shot across the bow as I depart. It leaves a memorable impression on those who experience such an Alfred Hitchcock moment, and proved to be available fodder for listeners on the show. Here in Grand Rapids, one such nest near the river and museum in a heavy foot traffic area, became such an exhibit to observers. During the nesting season in May and June, the comedic routine never gets old to bystanders, and is downright startling to victims. This is a bird that never heard of or acknowledged the old adage "pick on someone your own size."

The common blackbird, a loosely-used term for different species of rather unremarkable non-colorful birds and the grackle (*Quiscalus quiscula*) with their purplish and black sheen, became a subject one morning on the show. Stanley from Standale called to report trouble with his community birdbath. He enjoyed watching birds lounging at his bird bath and splashing in the pool. It can be quite entertaining and colorful to watch songbirds hanging out at the pool having a drink. Recently, however, frequent visits from a "blackbird" with something in its mouth had changed all that. His question made for humorous soiling of the airwaves.

It seems the grackle was quite disciplined in cleaning the nest of the odorous discharge of the clan. Can you blame him? I would too. They clean the nest to throw off predators from the scent, "flushing" the nest by finding water in close proximity and carrying the offending droppings there to make a deposit. Now, a bird bath is better than the family swimming pool. Feathered friends and popular socialites, like the cardinals, had decided to avoid Stanley's "pool." They had put his lounge out of business, due to the undesirables hanging out at his joint and their nasty behavior. Word gets around and birds are adept at tweeting information to their friends. I had suggested Stanley put up a plastic chain with a sign "closed for cleaning." (Like the signs you find at a roadside rest area, usually when you really have to go.) Stanley questioned my advice, not believing grackles could read. Smart guy that Stanley. Instead, we opted to clean the bath and turn it upside down for the few weeks of the nesting season, while this bird did its bathroom spring cleaning.

It appears that birds are quite protective and tidy about their households. This was confirmed for me, when one of our listeners shared with me the cigarette habits of birds. No, they don't take a smoke break while foraging for worms. They don't even go out back behind the bird house to sneak a smoke. They don't have a two "peck" a day habit. The listener shared with me that scientists have discovered they use cigarette butts, weaving them into their nests.

Our urban neighborhoods are full of materials for birds to line their nests, from lint on the dryer vent to paper and, yes, even cigarette butts. Get off the expressway to the end of the exit ramp, and, while waiting for the light to change, try to count all the butts piled along the curb. It has long been assumed and observed that birds are resourceful, and use what the suburbs provide them. I watched a bird at an outdoor café fly back and forth, landing on an open table and taking one sugar packet at a time until the container was empty. Curious, I followed the bird's flight path and found it did not have a sweet tooth. It was lining a nest with sugar packets, assuming they would make great pillows I guess.

The results of the study shared with me, determined the birds were collecting cigarette butts by design. They somehow instinctively knew that nests with cigarette butts woven into their fabric, were less likely to contain bloodsucking parasites than nests that did not. Their nicotine habit, using smokes as building material, significantly reduced the amount of ticks present in the nesting material. I would expect puffin birds to find it hard to kick the habit. Next time you see a cardinal, finch or robin fly by with a cigarette in their beak, you'll know what they're doing. If you're not sure, just look for the Surgeon General warning imprinted on the side of the bird house.

One of my business associates had a unique experience, that caused for some uncomfortable radio moments for those having breakfast. She woke up during the night with what felt like water in her ear. Having spent time at the lake that week, she figured some moisture had accumulated in her ear. She got up and went to the bathroom to use a Q-tip. At this point, a stinging sensation, or bite, occurred, causing her to wake her husband. Pulling on her ear lobe to open her ear canal, he spotted the culprit, a large bug. They removed a disoriented earwig from her ear. This is not an old wives tale, earwigs do crawl into ears. In her case, it resulted in a red and swollen ear, and an ear infection requiring antibiotics. Now, it is unusual for an earwig to "wiggle" into someone's ear, but the bad news is that it does happen. The good news is they do not burrow into your brain; they are not inclined to see what is on your mind. Earwigs are creepy, and just like damp, dark hiding places. They have a bad reputation, but essentially are interested in moist, rotting wood, bark mulch and vegetation. Their "bark" is worse than their bite, but they look frightening with the pincers on the back of their abdomens. When they do get indoors, they tend to gravitate to places where there is water: such as bathrooms, kitchens or laundry. Eliminating their hiding spots is key, and identifying entry points into the house will solve the problem. Avoid mulch piled against the foundation, and set irrigation so it doesn't make foundations continually damp. Irrigation heads are for turf, not landscape plants, so set accordingly. Try to establish a "dry zone" along the foundation of the house of at least six inches. If control is necessary, some diatomaceous earth can be a good, natural solution. Make sure gutters and downspouts are functioning properly and drain away from the foundation of the house.

Take the same approach with slugs to avoid a slugfest in your landscape. Slugs can make swiss cheese of the foliage on your plants, and have a voracious appetite just like the deer for your hostas. Hosta la vista baby, if you don't take precautions. Having a soft-bodied underside and liking damp, cool hiding places, remove the mulch from the base of your hosta plants. Make it rough on them, considering again some diatomaceous earth, or, better yet, cut strips of roofing shingles in a dark color. Some people insist on little butter dishes filled with beer sunk in the soil, so the lip of the container is flush with the soil surface. Yes, slugs like beer; they crawl into the dish, lose their car keys and die happy. You're better off drinking the beer yourself, which will greatly improve the appearance of your landscape, at least for the short duration.

Julie called the show one Saturday morning, and, I could tell by her giggle, she had a story to tell. Julie and her granddaughter were going downtown to see the Nutcracker one evening. They got dressed up for the occasion, and she went to the closet to pick out a purse to complete her ensemble. While at the Nutcracker she felt something crawling up her arm. It was the dreadful marmorated brown stink bug that moves into your house in fall whether invited or not. This stink bug had found a home in Julie's purse, and was about to make its debut.

The bug is not indigenous to the US, but has become a big nuisance in the past few years. With sucking mouth parts, it does sizeable economic damage to foliage and fruit crops. When disturbed, alarmed or crushed, it smells something like a combination of skunk, cilantro and a gym bag with a hint of garlic. It doesn't like winter, so like the box elder bug or the non native Asian ladybug try to find their way inside our homes in October. It beats a hollowed-out log or leaf litter, I'm sure. When they find a good home, they send out an odor that attracts or invites others to follow suit.

This nonplussed bug had decided a night out was in order, and was obviously aware that Tchaikovsky doesn't stink. I personally much prefer the 1812 overture, however, you have to give the bug credit for its cultural effort and attempted pas de deux with Julie. With a hankering for fruit, I would suspect the bug used Julie for transport, and had its eye on the sugar plum fairy. Julie flicked the bug away with a sweep of her hand, and I would have loved to see the look in the eyes of the unsuspecting patron who later felt that bug crawling up his pant leg. Better yet, the look of the surrounding others in that row, when the unsuspecting attendee commenced the initiation of the bugs natural odor.

CHATTER ON A TALK RADIO SHOW ABOUT GARDENING AND plants can quickly be dominated by the animal kingdom. Leading the list is moles, then deer and rabbits, closely followed by squirrels and chipmunks. Skunks and woodchucks represent the animal kingdom well too, when it comes to conversational fodder for the show.

Pepe' Le Pew is a frequent topic in both spring and fall when skunks tear up the turf looking for insects to eat. They dine on grubs affectionately, known as "lawn shrimp," and rip up the lawn like a novice golfer on a driving range with a nine iron. Sometimes they camp out and have babies under porches and decks, which always creates a fun family adventure. I learned a long time ago, that a simple flasher plug and incandescent light bulb, flashing off and on twenty-four-seven can be an effective means of convincing the campers to move; it will certainly entertain your kids. If anything, the nighttime shenanigans of skunks, and, the ensuing damage and duress for the homeowner, provides me fertile fodder to roll out a pun-rich opportunity for skunk talk....

"I find them "odor-able"

I'll be brief because I am a man of "phew" words

They are an animal of "distinktion"

Skunk control makes "scents"

I crossed a skunk with my cell phone...now the service stinks

A skunk wandered into my pond, he "stank" to the bottom

However, one backyard animal can light up the phone lines

as well as a mole infestation, and that is the much-maligned marsupial … the possum.

One night I had a possum in my window well, and discovered him with my flashlight cowering in the bottom of the well. Playing dead as they do, I decided to retrieve a board from my shed and strategically place it in the hole, so he could climb out when ready to do so. During the course of the night, he did climb out, I presume from boredom. Other than the ramp to freedom, I did nothing to entice him to come out. I decided to use this experience as a topic on the radio show, to throw something out there and then ride the wave of interest with our listeners.

Vanessa from Mt. Pleasant Michigan called to share she had a similar experience, but used a different approach. She placed a ramp in the window well like me, but prior to the ramp tossed some pizza in the hole. It appears, according to her, that possums like supreme pizza with the works. This got the possum's attention. Now, for the coup de grace ,she decided to sweeten the pot by baiting the end of the ramp with leftovers. Barbecued ribs and hot dogs were placed at the pinnacle of the ramps for nature to take its course. The possum took the bait, leaving the window well, choosing the ribs and leaving the hotdogs. It appears possums do have discriminating tastes. Other callers chimed in with stories of how they love black licorice, which they use as the key component to baiting a live trap. Others called about their interest in grapes. Grape vines growing on a pergola or arbor present an irresistible opportunity to binge. One caller said he heard a loud thump on his back deck. It appears the possum, tight roping the pergola covered in grape vines, had become drunk on the fermenting grapes he was gorging on. I have heard that about possums and how they love to wine and dine on grapes. In his case, the possum later got up and left, I'm sure with quite a hangover and headache, probably losing his car keys and smart phone in the process. Possums also love cat food, but I'm sure would opt for a drunken escapade on grapes if given the option.

Once you have possums, you are likely to see them hang around as they have a great memory for food sources.

Possums can be quite beneficial in the landscape. If you can get past their strange looks and habits, some people build nesting boxes to attract them to the yard. A possum is a tick vacuum and can consume thousands of ticks a week from your property. They also help control rodents like mice. It's the strange tail, beady eyes and toothy death grimace that creep out most people. One caller, however, shared the story of raising a nest of baby possums and naming them, having them live in the house for years. As he put it "Georgine" would curl up on the pillow at bedtime and sleep with the family. I'm still feeling bad about the one I nailed with my car late the other night.

In late winter and early spring, the phone lines are full of frustrated lawn owners once the cover of snow has cleared. The key to moles is to make your neighbor's lawn more hospitable than yours. You are not battling an army of moles, you are trying to convince a cantankerous, isolated, selfish rodent to find residence elsewhere, namely your neighbor's yard. You would be cantankerous too if your full name was "Moldywarp," which is derived from an archaic Germanic word for "earth thrower." The reality is they throw a lot of dirt as they tunnel; they can run backwards as fast as they can forward, can do somersaults and can even swim. More importantly, they eat as much as their body weight each day, 365 days a year. It is, however, completely untrue that they prefer guaca-moley and chips or pancakes and molasses. No book based on gardening can afford to ignore this trouble-making, subterranean mammal, so, never fear, he will reappear down the road when we address lawn and order.

PEOPLE LOVE TO TALK ABOUT SQUIRRELS. I SPENT $150 ON a squirrel proof pole for my bird feeders. It took less than 24 hours for them to figure it out. They destroyed the device and the feeders, and would balance on my investment, seemingly mocking me from their perch. These lovable little fuzzballs know how to frustrate suburban homeowners trying to feed the birds. That's why I enlisted the help of George Harrison. No, not the Beatles George Harrison, the one that lives in Wisconsin and wrote the book *Squirrel Wars*. George tells me he has plenty of phone calls at home from people believing he is THE George Harrison of Beatles fame.

It is an awful feeling driving down the road and hitting an indecisive squirrel. Flying down the road and watching him roll from the back of the car in your rear-view mirror. I talked to George about why squirrels do that middle of the road thing. You know, run into the middle of the road and then dance back and forth, trying to decide whether to go left or right. He tells me it is because squirrels are much smarter than deer. Deer decide to run out into the road and have no second thoughts. They operate on their own agenda and have no regrets at their own peril. Squirrels, however, are wily enough to recognize that once they are in the middle of the road, they may have not made the best of decisions. That's when you get the centerline dance they do, similar to when you run into someone in the grocery store and go left and right with them in an embarrassing dance in the soup aisle. Squirrels are smart enough to employ a life-changing, decision-making process in the middle of the road. You, as a human, have to decide whether to hit the brakes or swerve at your own peril. It is a cat-and-mouse game played every day in the middle of the road in your neighborhood.

Plants are amazing because they can create their own food … animals can't. So, animals eat plants much to the chagrin and expense of fledgling gardeners. You can't blame them. We should eat more plants … food is medicine. Some plants don't want to be eaten. They have thorns or fuzzy foliage or smell different. Some grow only in difficult-to-reach areas. What we do in our urban landscapes is set out a protected buffet. Some plants want to be eaten to propagate their species. Avian species are especially good at this as the seeds they consume exit their backsides softened and prepared for good germination.

We combat invaders like deer and rabbits with repellent sprays. You must be persistent as some repellents are only as effective as throwing the bottle at the offending species. Monica, my neighbor, sets out trays of bird seed and treats for the deer. I counted 16 of them in her front yard one night. She then sprays all her shrubs with Liquid Fence deer repellent before going inside for the evening. The deer dine on her seed buffet, then move to the neighbors for the salad bar and/or dessert. It's hard to get mad at a wobbly-legged Bambi, but they do a lot of damage in short order. The deer have become a cause celebre in my neighborhood, and seem to have the upper hand.

The deer in my neighborhood have that look in their eyes like, "what's your problem" when you get close. Urban living is easy, and they have no intention of returning to their rural roots. I had an experience running on a paved bike trail while training one evening. As I ran the trail, off in the distance, I saw three deer lounging on the trail. I thought, this is interesting, I wonder how this is going to turn out. As I closed on their position, they remained on the trail, barely acknowledging my existence. Their laissez faire attitude, similar to what you would expect from a cat, surprised me. As I got within feet of their resting place, they stood and began moving in the same direction as me. I continued along, realizing we were running side by side along the trail. I would have loved to have my phone with me to videotape the event, because we ran together for what seemed like a minute. *No one would believe me,* I thought, as I had my Grizzly Adams one-with-nature moment, galloping with the herd. Just like that … they were gone. After they made a sharp, 90-degree angle, bounding into the woods, I questioned what I had just experienced. I chuckled, realizing I hadn't passed the buck; it was an experience better than watching the National Geographic channel.

I have had great success coexisting with deer by planting alliums, hellebore, ornamental grasses, nepeta, lambs ear and other deer-resistant plants. Not tonight deer. Many books have been written and lists generated, simply a search away, for the frustrated home gardener. The tough part is you may have to abstain from certain cherished plants like tulips, that deer and other herbivores simply find irresistible. With the European blood running through my veins, I find it hard to kick my tulip habit, so I resort to spraying repellents, from the time they pop their head out of the ground, until they have completed blooming.

I have had success by incorporating deer-browsing-resistant plants. Lists of regionally resistant plants are abundant on the internet. Just remember, if hungry enough, herbivores like deer and rabbits will eat virtually anything. I've seen them eat a stringy yucca in the dead of winter. Sure, you can build a wall or surround your property with a tall fence, eventually making it look like a gated compound in Beverly Hills. You can have dogs patrol the property, and they are effective in deer control, but remember, not all dogs are garden friendly, and they are excellent diggers. Dogscaping is a garden trend where we coexist with our canine friends, providing distractions, digging areas and running routes.

The lengths to which people will go to coexist with deer can get rather humorous. When you are willing to live with hair-stuffed panty hose, hanging around the yard enhanced with soap shavings from a cheese grater, you're engaged in some rather serious deer diplomacy. Aluminum pie pans strung on wires and sweaty clothing strewn on shrubbery is enough to give your neighbors concern. When you take it to the strobe light and loud music with motion-activated sprinklers level, you are apt to have the police knock on your door. Predator urine, blood meal, rotten eggs and open, wet bags of organic sewage fertilizer, with a dash of garlic and red pepper might work for a while, but you've sur-rendered your surroundings to the beast. The panty hose thing is just a little too kinky for the average neighborhood. Maybe I'll just stick some pink flamingos in the lawn and go inside. I'll deal with it another day. Not tonight, deer, I've got a headache.

An impatient man wanted it faster
He ignored the weather forecaster
He then slipped on the mud
And fell with a thud
Right on his beloved Aster

Chapter Six

Operating by the seat
of your plants

WHY IS IT THAT MANY SUCCESS-
ful or famous men and women came from
troubled homes or difficult childhoods? Do
they draw strength from their hardships and grow from it, or
are they simply marked for greatness? Is it resilience, survival,
motivation … or all three?

In the plant world, I find a ponytail palm to be innately resilient, able to rebound from periods of stress due to its inflated survivalist camel-like trunk, conserving for a better day. Plants like
lithops, aptly named "living stones," are survivors. Flowering
annuals like portulaca are resilient in the face of extended periods of drought or heat. Many flowering annuals, if given some
duress, and water is withheld, fight back by blooming better than
ever before for seed production. Some water stress gives them a
kick in the plants, better than any fertilizer, to ensure that future
generations develop in the form of their flowers and seed production.

IF THE POET DYLAN THOMAS WAS RIGHT WHEN HE SAID "the only thing worse than an unhappy childhood is having a too-happy childhood," then he would have appreciated some flowering impatiens, or, as one of my gardening mentors called them "busy lizzies." Don't pamper them to get them to bloom. A little bit of stress will give them a kick in the plants.

A plant, just like a person, can decide when told, "It can't be done" to choose from the following options:

- I'm going to give up.
- I'm going to get angry and pout.
- I'm going to prove them wrong and succeed.

Those who experience some adversity, seem to be higher functioning or satisfied, than those who have had no adversity at all. It explains the life of an epiphyte like an orchid. They cling to something to thrive with a survivalist attitude, not a parasite that drags everyone down, but a presence that hangs in there and thrives, blooming in the face of a challenging environment without complaining. They are better for it. When they have a good grip on their station in life, the roots send a signal to the plant to bloom its head off.

I like how rhododendrons mope in winter with drooping foliage. Yet, there parked above the fussy foliage, is the promise of spring in the form of a bud mounted on top. They eventually get over the winter weather and achieve their potential when the foliage lifts to celebrate the breaking bud and expanding blooms in the warmth of May and June weather.

Some pruning is to be expected in our lives, whether it comes early in life or later. Besides, pruning is a naturally instinctive inclination for humans who want some measure of control over their prospects. We want to be involved in the process, to be part of the story line. The general population loves to encourage a rising star to achieve celebrity status, only to tear it down, then embrace and celebrate a comeback. It's a story that has played out in

Hollywood, sports, conflict and politics repeatedly, through the ages and the plot to many great movies. The star (your plant) has grown out of control, and, just when it appears all is lost, you, the wise sage (with pruning shears in hand) put the plant in its place. The setback (pruning) brings order to the world (control of your landscape) and we celebrate the star's rejuvenation with a happy story ending. We get a shiny trophy as best supporting actor.

FRANKLIN DELANO ROOSEVELT MADE A STATEMENT AT A commencement address at Oglethorpe University in May of 1932. It is understandable why he made it. Unemployment was at an astounding 24 percent, and not just industrial business; it was also the "dirty thirties," where drought and farming practices created a dust bowl, affecting millions of acres. As president, it must have been a difficult and dark time for him as the troubles mounted, including those developing in Europe. To go from unemployment rates of around four percent in 1929, to about one quarter of the populace out of work must have felt like the sky was falling. What do you say to graduating students in that environment? He said something that I keep posted on the wall of my office. I refer to it often.

> Do something.
> If it works, do more of it. If it doesn't do something else.

> *"It is common sense to take a method and try it. If it fails, admit it frankly and try another. But above all, try something.*
> *–Franklin Delano Roosevelt*

Others have attributed that quote to Roosevelt as, "Do something. If it works, do more of it. If it doesn't do something else." I believe many people buy a plant not knowing where it will go. When you get home, you walk the yard with the plant in one hand and a shovel in the other. Admit it, you've done it, the impulse-plant-buy. I do it all the time. You might not have a plan, you might be operating by the seat of your plants, but you're doing something. Something isn't going to grow if you don't plant it, right? Weeds will grow, but you changed the course of history by planting something else.

Warren Buffett the famous business magnate, investor and philanthropist is attributed with a quote that says, "Risk comes from not knowing what you're doing." That's true when investing money, building a house, flying a helicopter or running a chainsaw. It is also true, however, that without taking some risks, you might end up doing nothing. And nothing doesn't get you very far. You must sow some seeds in faith to cultivate some growth.

ARE YOU A GARDENER OR A "GARNERER?" THERE IS A DIFference between the two. A "garnerer" is someone who garners, collects and gathers plants. A "garnerer" can collect praise for their massive collection of plants. We've all seen the home in the neighborhood with every possible space consumed by their collection. With no rhyme or reason, the scope is a trompe l'oel effect, rivaling the gardens of Pompeii after their volcanic demise. With pride, the good, the bad and the ugly are on display for all to see in amalgamated assortment, much to the chagrin of the neighborhood association. Conversely, a gardener understands that plants have a purpose. Their placement is with purpose, both for the welfare of the plant and the eye or consumption of the beholder.

When it comes to words as gardeners, we have a guy named Carolus Linnaeus to thank, or blame, depending on the side of the fence you stand on. Mr. Linnaeus decided he was going to take it upon himself to be the father of taxonomy. We use the system he developed in the 1700s to classify and organize living things like plants. One of his biggest contributions was the two-part naming system called binomial nomenclature. For example, you just might have a *Metasequoia glyptostroboides* growing in your yard and not realize it. If you do, and can pronounce that botanical name, it will impress neighbors and family as it rolls off your tongue. The common name of dawn redwood is much more user friendly.

The tough part of his system is that he picked Latin to organize the kingdoms, classes, genus and species of plants even though he was Swedish. The rumor is he chose Latin because it is a "dead language." If he had chosen English or Dutch or Vietnamese, then people in other countries would be offended that their country wasn't chosen, and would choose to have no interest in plants. (I made that part up.) The Greek language also played a significant role at that time, so that's why plant names for many people are Greek to me.

Once you learn the Latin names they can make some sort of sense. When a plant, for example, has *Canadensis* in the botanical name, odds are it is native north of the border. In some cases, the name, like a favorite perennial of mine *Pulmonaria lungwort*, suggests the plant was used for respiratory medicinal reasons. Or, when you see the word *tomentose* in a plant description, the leaf or stems are hairy or woolly.

There was and is a method to the madness. The system, however, is intimidating to the average person who passes at an attempt to verbalize a plant name, due to the risk of embarrassment. Once you master the pronunciations, they can make you sound really smart and advance your opportunities for consideration to an important job. For the weekend "yardener," however, the intimidation and embarrassment of some of these names are too much to handle. I love the confidence of some people who are more than willing to butcher the name in the sake of getting what they want. You get the general idea. If you have a shopping list with *Ranunculus*, *Lisianthus*, *Ligularia*, *Nierembergia* and a *Bougainvillea* to boot ,you have to go into it with a degree of confidence, whether you know what you're doing or not. Some of my favorites through the years with variable, creative inter-pretations as an example are:

Thuja, pronounced Thu-Yah
Cotoneaster, pronounced Cot-Own-EE-Aster
Syringa (lilac) pronounced Si-Ring-Ah
Tsuga (Hemlock) pronounced Soo-Gah

I have clobbered my share of names too through the years. It's always great for a chuckle to discuss clematis on live TV, you call it klema-tis and I call it clem-ah-tus. Or shoppers looking for the flowering shrub weigela (pronounced wi-gee-lah) who either insist on adding an "l" at the end, making them "weigelia" or giving up all together and asking for those "weegly" things. Good thing *Galium aparine* is considered a weed and unde-sirable, or people would be asking for it by its common name "sticky willy."

ONCE YOUR PLANT IS NAMED IT NEEDS A HOME. LIKE YOGI Berra said, "If you don't know where you're going, you're going to end up someplace else." Without some kind of road map, the placement of your plant material will be an adventure. Consider alone that this is foreign territory to someone who doesn't do this every day. And do it yourself bravado is evident by the thousands of garden accidents, sending people to the hospital or doctor every year.

When my son was young, he wanted to cut the grass and run the power mower. I was more than happy to turn over the monotonous drone and task of cutting the grass, but feared for his safety. After training and educating him, I let him loose just outside the reach of the long arm of the "lawn." He would cut the grass, but, from time to time, become distracted or stop to tinker with the mower engine. Frustrated parents say stupid things. I found myself popping my head out the back door hollering at him, "If you cut your legs off don't come running to me!"

Maybe so many accidents happen in gardens, because they are a place people want to relax in. I have learned that a fall is the most common accident, and a cut is the most threatening, followed by being struck by things. Projectiles from a lawn mower or the rake that gets stepped on, and the resultant bump on the head, would be funny if it wasn't so dangerous. Of course, as we've found on the show through the years, there is a direct correlation between the consumption of adult beverages and number of incidents in the garden. A can of beer in one hand, and a pitch fork in the other while mole hunting, is not a good combination.

It didn't surprise me that lawn mowers cause the most serious yard accidents, but … flower pots are dangerous? It must be the trips and stumbles over a pot, or the lifting of pots, that put it high on the list for back injuries. Or maybe it's the combustible pots in the heat of summer. There are stories of peat that spontaneously combust in the sun. Or the smoker who puts his cigarette butt in the window box that is left to smolder. When it comes to danger, I read a study that said lawn mowers top the list, followed by pots, pruners, spades, shovels, electric hedge trimmers, shears, pitch forks, plant supports and believe it or not garden hoses.

Garden hoses just look so innocent but are a snake in the grass. It's fun to see people set their oscillating sprinkler, the type with two settings, drip and monsoon. They run from the sprinkler to avoid getting wet, and tripping on the hose in the process. You know there must be an open screen window nearby, usually your neighbor's. It's amazing that garden hoses and sprinklers cause a large number of injuries annually. I heard an estimate that far surpassed amusement park rides. Yogi was right, if you don't know where you're going, you're going to end up someplace else … the doctor's office.

I suppose live radio is dangerous. You are always seven seconds (delay) from humiliation. When taking a call on the air, if someone leaves their radio on, you can hear yourself seven seconds ago.

Some people simply do not know how to be interviewed. It drives me crazy, especially sports interviews, where the interviewee repeatedly uses the phrases "Um" and "You know." I understand this can be nerves, however, if the phrase "you know" is repeatedly used after each expression, I want to say the caller, "If we knew, we wouldn't have asked the question in the first place would we?" Others fill the space or pauses with the word "Um." Um is defined in the dictionary as hesitation or I do not know. So, when you Um all the time does that mean you are stalling to make it up as you go? Or you're uncomfortable with hesitation, pauses and pacing? When you do live radio, you become sensitive to this issue. If you listen to recordings or podcasts of yourself, you train yourself on the delivery. That spacing is also needed in the landscape. To be good with spacing … air. Know the eventual size of the plant, not its current size. This mistake is made repeatedly in home landscaping.

I FELL OUT OF A TREE WITH A RUNNING CHAINSAW ONCE. I was young, adventurous, indestructible and stupid. As you grow older your mind becomes seasoned and wise like an old tree. I have learned to be humble or be humbled. It does pay, however, to be lucky from time to time. Unless you try something beyond what you have already mastered, you will never grow. The key is to approach the something new with humility, common sense, and thought and respect for a process. When you're young you tend to jump in with both feet. Act first ask questions later. Today, I look back and am amazed I'm still alive to talk about it. I have conversations with a big, old wise oak tree along a walk I frequently take. We discuss the mistakes I've made in my life. I do most of the talking … he's a good listener.

The tree I fell out of was a big, old sugar maple tree. A majestic, huge and weathered tree that came in sweetly in spring and went out with a bang in fall. Orange and yellow with hints of red, when lit up in October, was a sight to see. The tree had some old branches on the interior in decline that I decided to remove. I was as broke as a limb in a windstorm, so I decided to save money and make it a do-it-yourself project.

I leaned a ladder against the tree to allow me to climb to the lowest limb. With chainsaw in hand, I climbed the ladder and stepped out on the limb. I believe in life you have to go out on a limb, because that is where the fruit is. Just don't go out there with a chainsaw. Leaving the safety of the ladder, I now proceeded to climb up the tree, limb by limb, with chainsaw in hand.

About 25 feet up I reached the target of my ambitions, an old crossing limb on the interior that I had been eyeing for some time. I positioned a foot in a crotch, between the trunk and a limb to steady myself, and lifted the chainsaw on a branch to begin. Pulling and tugging on the cord, the chainsaw finally roared to life. In a testosterone-driven moment of insanity, I stood tall and stupid in that tree, master of my domain.

What happened next … I do not know. All I remember is being on the ground with the chainsaw still running about 3 or 4 feet away from me. Your impulse is to jump up and quickly look around to see if anyone else saw what happened. Pain and injury impulsively play second fiddle to embarrassment. I dusted myself off and hit the kill switch on the chainsaw. I decided that old limb had been around for 50 years or more, and, if I wanted to live to that age, I should just walk away and live another day. I would go do something safer like groom the petunias in the front yard. My Paul Bunyan impulses would soon enough return, and maybe today wasn't the day to act on them. Sure enough, opportunity knocked and within weeks and I was back at it again.

I ARRIVED HOME FROM A LONG DAY AT WORK AND MY KIDS were visibly excited. Mom had something exciting to tell me, and they could hardly wait for me to get home. The neighbors across the street, Marv and Nelvina, had decided they were going to gift their above-ground pool to us, because their kids were full grown and leaving the nest. My kids, who were young, would inherit the pool. All I had to do was move it from their yard to mine, and let the fun begin. Expecting me to jump up and down with excitement, you can imagine the family's disappointment, when I quietly hung my head and closed my eyes. I had to execute a plan and follow through, or there was no hope to be Dad of the year and recipient of a World's Best Dad coffee mug on Father's day. What could go wrong? It was an aging, 24-foot-wide vessel of thin steel and aluminum frame, with liner, held together by water pressure and a few nuts and bolts. I was captain of my destiny, and, to my kids, I was as good as Jacques Cousteau when it came to water adventures. In the spirit of Admiral David Farragut, I had no choice but to proceed, and it was "Damn the torpedoes full speed ahead!"

One of the first tasks at hand to prepare the designated site for the pool, was to remove some large, overgrown trees on the property. One of those trees was a massive, tall cottonwood tree. One could hug the trunk of this behemoth, only beginning to wrap their arms part way around its impressive circumference. Its limbs blocked the sun like billowing, threatening storm clouds. This tree, the scourge of the neighborhood, had plugged air conditioning units with copious amounts of cottonwood fuzz for years. The tree had met its match ... me and my chainsaw. To be captain of my ship I would have to first conquer the land. The gauntlet was thrown down, and I was picking it up to accept the challenge, understanding no one was aware of my sugar maple faux pas weeks earlier.

On a blistering, hot, summer day, the fuel in the gas tank of the chain saw bubbled and boiled in its labors seemingly troubled by the fact that I had no idea what I was doing. Today the cottonwood tree was coming down. Understanding that the tree would have to be notched at the base in the direction I wanted it to fall, I approached my work like a Jedi knight with light saber in hand. The teeth of the chainsaw spewed chips around my feet, and the engine groaned at the task. This tree was a monster, and I was way out of my league. Part way into the tree, a slow and nagging realization of the scope of the project dawned on me. Having notched the tree, I shut off the chain saw and looked at my neighbor's garage within striking distance of the tree. I needed a backup plan and came up with a strategy.

I had bought my brother in-law's old Chevrolet Monte Carlo as our family transportation and decided I would test its versatility. I drove it into an adjoining field and proceeded to tie ropes to the bumper of the car. I then leaned an extension ladder to the trunk of the cottonwood tree and tied the other end of the ropes to the trunk of the tree. Climbing down the ladder and into the car, I put the car into drive and took the slack out of the rope. Giving the rope a gentle tug with the accelerator, I was successful in pulling the bumper off the family car. Now I was stuck. The tree was leaning, and I didn't have the money to replace my neighbors garage. I didn't want to know what was in my neighbor's garage. I prayed for a miracle. Moments later it arrived in the form of an old four-wheel drive Jeep driven by a man named Otto. I flagged him down in the road and breathlessly pleaded my case. There was no time to waste. We transferred the ropes from the wreckage of my General Motors '70s model of ingenuity to the hitch on his Jeep. He gassed the Jeep while I cut and miraculously the tree fell in the general proximity I needed it to with a massive crack and thump. I offered a prayer of thanks and gave Otto the last ten dollars in my wallet. Disaster had been averted … at least for today.

Today, every time I hear a chainsaw run, I think of that old cottonwood tree. It reminds me of the importance of tree pruning, especially for trees that are near structures. It reminds me that trained professionals should cut down trees and that they are dangerous. Having watched my son climb around trees with rope and harness and drop trees with precision in a prescribed spot, I always recommend calling a professional arborist.

Having cut down the trees, and having prepared the site, I began the task of assembling the pool. Piece by piece and floppy, rusty, flimsy-metal-panel by metal panel, the circular structure soon took shape. Working quickly on a weekend, I was bolting and fabricating my makeshift used swimming pool, hoping to fill it with water soon. The kids were watching from inside the house with their faces pressed against the window. For that reason, I was out there on a Sunday afternoon tightening bolts and attaching parts by the seat of my pants, when my friend Hank stopped by to caution me. Like John the Baptist, he told me my pool would not be blessed, because I was working on it on the Sabbath. I felt more like Noah and I was on a mission. Sure enough, that night the storm clouds gathered, and a wind storm took down what man's hands had wrought. With no water pressure in the pool, the panels were helpless in the wind and were trashed in a pile in the side yard. The prophet had spoken in my wilderness, and his prophecy had come to fruition. I would have to start over.

> Your landscape is like your life. The objective is not to "prove" yourself. Your intent should be to "improve" yourself.

You must learn from the mistakes of others. You can't possibly live long enough to make them all yourself. For many, the exercise of gardening is curb appeal. How I look to my neighbors … a horticultural selfie for all to see socially. Here is where the landscape at our front steps and go off the rails. You can opt to plant the same thing everyone else is in your neighborhood, to follow the crowd and fit in, or, you can break the cycle of cemented mindsets and go out on a limb. The important thing to remember is the following. Your landscape is like your life. The objective is not to "prove" yourself. Your intent should be to "improve" yourself.

IT IS PROJECTED THAT, IN THE COMING YEARS, THE PERcentage of city dwellers will increase from 54 percent to 66 percent of the world's population. A move from the rural to the accessibility of city life. Older generations like the convenience of walking distance, and younger generations like the opportunities for experience and lifestyle. This move will change how we enjoy our landscapes or use plant material in tighter or smaller spaces. It has also produced a resurgence in indoor foliage plants for "breathing" rooms.

Container gardening also enjoys great popularity due to the move to the city. When planting a container, focus on the center of your planting first, and work your way to the outside. Have a basic understanding of the plant's eventual size and habit. The brains of the operation is the focal or thriller plant in the center of the pot. The gravitas comes from the fillers and edgers. The spillers or trailers can't contain their excitement, overflowing the sides. Thriller Filler Spiller. Or look at it as a planting progression of Focal, Filler, Edger, Trailer working from the center to the perimeter. You don't have to "pot" up with a boring container.

Plants will always be inspiring, and, after years of walking around with a plant in one hand and a shovel in the other, trying to find a place to plant it, I have learned ten-easy-to follow rules to a better home landscape design. Walking and running through neighborhoods, it is easy to see who had a plan, who had a concept, and who was operating by the seat of their plants.

1. If you "over do" it, you will have a do-over on your hands. Use the look-around rule. Avoid what we call monoculture. Too much of one thing can be a problem. Diversity is important in the landscape. If you plant too much of one type of plant, and a problem crops up like an insect or disease, it will spread like wildfire without a proper amount of diversity in your yard. Look around your yard, your neighbor's yard,

is there a plant that is let's say, over done? Approach your landscape in bite-size pieces. It's more fun that way and gives you opportunity to change your mind as you go.

2. It is possible that opposites initially attract in relationships, but over time they will drive each other nuts. People tend to gravitate toward those of similar tastes and characteristics. Unlike people, plants don't walk or talk, so they have a natural understanding of their differences in the landscape. I know that in the landscape, opposites "do" attract and make for a better look. Use opposites in color, texture and form. For example, the colors blue and yellow are opposites on the color wheel. Together in the landscape they make for stunning partners. Another example would be form. If every plant had small leaves, the look would quickly become confusing and cluttered. Use of large-leafed plants, with small-leafed plants, allows the plants to show off their unique characteristics as part of the whole presentation. The landscape will be less busy to the eye and more focused. Each character gets to play a role. Ornamental grasses are perfect for adding foliage differentiation to deciduous flowering shrubs or large-leafed perennials.

3. The law of similarity would imply that similar things tend to be grouped together. The individuals become a group, visually, as opposed to a collection of individuals. To show off a focal point for the wow factor, the supporting group needs to complement the star of the show by NOT having similar form or size and texture. It's like the supporting cast, showing up in the same dress as the diva in the performance. It's not a good idea and will ruffle some feathers. In the landscape,

the focal point (diva) is usually the plant you shelled out a lot of money for at the garden center.

4. View your landscape as though you are taking a picture. Visual composition understands "Positive space" (the main focus) and "Negative space" (the background). Negative space is not meaningless. Negative space supports the foreground or positive space in the picture. Positive and negative space together tell a story just like your landscape. Each have a role to play. Focal point, or wow plants, are supported by foundation plantings, often evergreens that provide the bones and structure of a well-designed landscape. Plants, like people, consist of movie stars and those who are content to play a supporting role. It's just the way it is in life.

5. Use unequal, not measured spacing. Reserve your tape measure for building stud walls or installing cabinets. However, in most cases, plants in a row with equal measured spacing do not look natural.

6. After you apply the above rules; add the rule of planting in odd numbers. Generally, plants in groupings of 1,3,5,7, etc., look better than even-numbered groupings. A good example would be the ubiquitous practice of two identical plants on either side of the front entrance steps or opposite corners of the house. I think the person who started this was the same person who started the ill-advised "volcano mulching" around trees. A misdirected practice that for some reason was copied by others as a generally accepted practice. How boring.

7. Use curved borders or edging, and tie it all together with a continuous border. A recent study I read sug-

gested that curved beds alone could add to the value of your house by one or two percent. If you are talking about a $200,000 home, that's $2,000 to $4,000 dollars! Once you've established the curved borders, consider a plant material that can tie it all together. The law of continuity proves that, points connected by curving lines are seen in a way that follows the smoothest path. The flow is seen as belonging together. Remember point 1, above, where we strive for diversity in plant material in the landscape? Now that you have a diverse group of plants, pick one type of low-growing plant for the border to tie it all together. It works!

8. Ask yourself (on paper) a lot of questions before you start, as it regards to your intentions for the landscape. What is the purpose of my landscape or this area of my landscape? What is my favorite garden use? Entertaining? Tanning? Bird Refuge? Bocce ball? Impress the neighbors? What is my favorite garden mood? Seclusion? Natural and unkempt? Bright and happy? Shady and mysterious? What is my favorite sensory effect, sight, and smell, hearing, taste or touch? How about favorite garden feature? Fountain? Pond? Path? Statue? Furniture or favorite chair? Specimen plant? By asking these questions and putting them on paper, you can do some goal setting for your landscape.

9. In the Midwest and Northeast of the country, the side of the house you plant on and the winter exposure to the plants, is an important consideration. I find the east side of the house to be a great area for planting. Plants like roses and hydrangeas seem to do well on the east side. The south and west side of the house

provide plenty of light in summer, but can be harsh, especially in winter to evergreens and broadleaf evergreens. The unforgiving structural shade of the north side of the house can be difficult for plants that need a little more light. Take all four seasons, including winter, into consideration when selecting the exposure for your plants. Plants susceptible to winter sun and wind, like broadleaf evergreens (rhododendrons) or susceptible deciduous (*Hydrangea macrophylla* or Japanese maple) may prefer the cover of a north or east exposure. Be attentive to natural microclimates in your landscape, and plant species accordingly. Some plants thrive in my landscape with the protection of the north side of a structure (when the sun is low in the south in winter) or the east side of the home where the snow piles in winter (natural insulation) due to prevailing winds from the west.

10. Put a shovel in the ground! Working in organic material for good drainage and a good ratio of oxygen-to-moisture retention capability is so important. Mix amendments in with the parent or existing soil fifty-fifty. Dig a hole in the soil where you intend to plant, and fill the hole with water. Observe what happens. Does it drain quickly? Slowly? Doesn't drain at all? This simple test will tell you a lot about what you need to do to make the planting area a good habitat for the plant. Composted manure, compost, leaves, peat moss and other organic material, worked in liberally with the existing soil in the planting area, will allow your landscape to thrive. What was it the presidential candidate Ross Perot always said in his campaign? "Measure twice cut once." A famous Abraham Lincoln quote was, "If given six hours to cut down a tree, I will

spend the first four hours sharpening the axe." When working with plants, look down first, not up. A good foundation is a recipe for success.

11. Consider the eventual plant size. Don't put a plant within two feet of another that will eventually grow to five feet! Do your homework. Take the mature, estimated size of both plants and divide by two. This will give you proper spacing between the plants for future growth. If you don't, soon you'll be moving or pruning one or both plants! It is less important with perennials, as most of my perennials are eligible for frequent "flower" miles they travel so often. That's half the fun of perennials, being able to bend over and split your plants. With trees, evergreens and shrubs, however, a bad placement can create dilemmas down the road. Don't let a poor placement eventually "shrub" you the wrong way.

12. Design the garden to be viewed from the **inside as well as the outside.** Grow a diverse group of plants, as well as plants that tend to give three to four seasons of interest. This will give you something to enjoy year-round, from your vantage point, whether inside or outside the home. Look at your landscape from inside the house before putting a shovel in the ground. A landscape should not be installed simply for curb appeal. Enjoy it from that favorite window in your home. Select some plants that work hard for you, providing three to four seasons of interest. As an example, I have always tried to incorporate an oak leaf hydrangea. Pretty, oak-leaf-shaped foliage in spring, gorgeous panicles of bloom in summer, stunning dark burgundy color in fall, and cinnamon exfoliating bark in winter. Finally, remember it is

okay to kill some plants. If you haven't killed any plants, you're not trying hard enough. Trial and error, or, as I say on the radio show, "trowel" and error. I have learned through the years that only a small percentage of people in life use Plan A. A far larger percentage end up using plan B, C or D. Instead of resigning yourself to the cards you're dealt, I believe often motivation can trump intelligence. G.K. Chesterton, an English writer, was known as the "prince of paradox." He is quoted as saying, "One sees great things from the valley, only small things from the peak." Remember, it is natural that when you make some mistakes, you will see and learn from them. Stay grounded my friends.

His prospects were looking grim
And his options becoming quite slim
He followed his gut
Climbed out of the rut
Life's more exciting out on a limb

Chapter Seven

May the Forest be with You

I WALKED OUT OF THE HOTEL INTO THE cool evening air, enjoying the crisp feel of the October night air in Cleveland. Wearing shorts and a tech shirt with running shoes, it was the perfect night for a run. Traffic rushed by me, and the tall buildings echoed the sounds of a busy city, bouncing off the walls. The concrete environment reflected the energy as I tried to orient my location and course. I looked into the night sky dotted with stars and chose my course. My direction would be impulsive, without planning. Breathe the air, run the pavement, enjoy the sights and sounds of a bustling city.

After running for about 20 to 30 minutes, I sensed the environment changing around me. The leaves on the sidewalk trees were brown and rustled in the breeze, while the shadows were long on the dimly lit street. Steam rose from the gutter grates along the road, giving a mystic and eerie sense to the concrete and asphalt landscape that now surrounded me.

I continued my journey, but, sensing I was not alone, I glanced to the side. A blinking light caught my attention as I was running. It moved with me as I continued my pace through an urban neighborhood that seemed strangely quiet. Lost in my thoughts, curiosity now got the better of me. I slowed to a stop, standing at the edge of curb and spotted him. A large man on a bike, wearing a helmet and shadowing my movement. *This is strange*, I thought, until I saw the decals on his bike, clothing and helmet. It was a police officer on a bike braking next to me on the curb.

A large man, he straddled his bike and looked at me saying, "You're not from around here are you?"

"No" I responded.

In my surprise I blurted out, "Are you pulling me over?"

He grinned looking down at me from his perch and pointed in the opposite direction.

"You want to be running that way" he said.

"Oh good," I said, "For a second there I thought I was speeding."

He hopped back on his bike and disappeared in the shadows of the night. I stood there for a moment, surprised by what had happened. Never in my life had I been pulled over by a police officer while running. I decided to take his advice, looking over my shoulder as I ran. I was relieved when making it back to the hotel. The police officer had changed my impulsive direction. I was in the dark and glad he hadn't asked for my license and registration. You never know when someone will come along, out of the blue, and bring about a course direction.

We often take for granted the beauty that surrounds us. We were reminded of the importance of diversity when the emerald ash borer (EAB) wiped out the ash trees lining entire streets in our neighborhoods. You know what they say about the suburbs, a place where they cut down all the trees and then name the streets after them. Trees teach us valuable lessons of diversity and perseverance and value. In our landscapes, planting a diverse grouping of plants is far healthier than to focus on just one species or variety of plant. Ash trees provided great shade and a great neighborhood canopy, but, when we forego diversity, we pay the price when a problem comes along. The problem spreads like wildfire. I embrace diversity both in my landscape and in life.

When a tree dies or is cut down, what we took for granted suddenly becomes obvious. It's like elevator music, you know it's there, but you're not really listening. It serves a purpose but is not a part of your thoughts until something happens.

When the iconic rock-and-roll legend Tom Petty died, it affected me personally and got my attention. Many of his songs played a role in the soundtrack of my life. As a teenager in the '70s, his songs just struck a chord with me. I was running down a dream. Petty was quoted in *Rolling Stone Magazine* about the possible end of his touring days as saying, "It's like you got a tree dying in the backyard. And you're kind of used to the idea it's dying. But then you look out there one day, and they cut it down. And you just can't imagine that beautiful tree isn't there anymore."

His quote made me think that the same rule applies to life. Often it is the hard working, enduring, low drama, high achievers in life that people take for granted. They just expect them to be there with their nose to the grindstone. In the music industry people come and go with lots of one-hit wonders. Petty and his band had decades of staying power, were loyal and hardworking, and, despite hardships and mistakes, which we all have, he got the job done. Is there a person or people in your life, who, from time to time, are taken for granted, and you can't imagine them not being there anymore? Petty was definitely an artist that marched to the beat of a different drummer, and he was good with that.

I LOVE THE GINKGO TREE BECAUSE IT MARCHES TO THE beat of its own drummer. The tree is one of legend and lore with a rich history. Even though it grows in many countries on earth, nowhere is it more revered than in China. With trees thousands of years old, it has been used for medicinal and healing reasons as well as their distinctiveness and longevity. It gave artistic inspiration and symbolism for cultures throughout the ages. Today, it is one of the world's most popular street side trees, because of its durability, size and low maintenance. Many artistic drawings and paintings have been done of its distinctive leaf. Taking it as a supplement is supposed to enhance my cognitive abilities. I have been the subject of good-natured ribbing for my incessant promotion of the tree. My favorite characteristic of the tree is that in the fall, the leaves tend to all drop at the same time. Lighting up the fall sky, the foliage turns from green to a bright, sunny yellow in fall. It litters the green turf with golden yellow splashes.

Organizations have held contests to guess the day the foliage will drop. I had a large gingko tree at one home I owned, and I loved it in the fall season. I would leave for work one morning, and, upon returning home at night, all the leaves would be laying at the base of the tree. Unlike the Maple or Sycamore, that would drift their foliage intermittently for days out on the lawn, the ginkgo, once it makes up its mind, it makes up its mind. This tree is not indecisive when a change is needed. For those of you who are frustrated by those who vacillate, this tree is for you.

EVERY YEAR IT SEEMS, WE EXPERIENCE A STORM EVENT where trees in our neighborhoods remind us of their presence, succumbing to a wind or ice event and causing damage. It is a natural physical reminder of "Entropy" in our lives. The definition of entropy is, in part, a lack of order or predictability; gradual decline into disorder or chaos. We all deal with entropy in our lives but prefer to ignore decline or deterioration. We work to move forward, because we are either moving forward or backwards … not much stays the same and certainly not for long. It's like dormancy, which is a reversible condition; you are either dormant or dead. Death, except for some notable rare occasions in history, is not reversible.

When it comes to trees, the entropic king in my opinion is the Colorado spruce. I've got the blue spruce blues. The ubiquitous, overplanted royalty of the neighborhood gets hammered by disease and decline. They look good for a while, and do a great job holding Christmas lights during the holidays, but decline (aka entropy) will soon set in. Cytospora cankers, rhizosphaera needlecast, phomopsis and needle drop on lower, sun-starved branches will soon leave you with a blue-green behemoth in decline. You can throw money at it, but sooner or later you will experience Blue Spruce burnout. I think it's why we see the resurgence of interest in the *Pinus* genus, fresh, fragrant and so North American ... a tree even Charlie Brown can love.

When it comes to entropy, we personally grow when we:

1. Recognize it as reality.

2. Develop the skill to identify it and resist apathy and complacency.

3. Understand entropy applies not only to the physical but mental too…our attitudes and relationships.

4. Most importantly, develop a clear understanding of what you can manage. Entropy dictates the need to manage what and how much we can manage in our realm.

This past year a major wind event took out a number of trees in our neighborhood. When they fell, it was a good reminder of necessary maintenance, planning and development to battle the ever-present effects of entropy. For example, to have strong trunks, tree trunks should be tapered from top to bottom. Trees develop tapered trunks from two major events in their lives: trunks swaying back and forth in the wind, and the presence of branches with leaves all along the trunk. These lessons are learned early in the life of a tree and its development. Without it, they won't be able to support the canopy of the tree when the storms of life arrive. Humans are like this too, the most intelligent have the ability to hold two opposing views in their mind and still be able to function.

The wind event reminded us of the importance of a tree's root collar. It is the area where the roots join the main stem or trunk. At the base of the tree there is a "flare" leading to the major roots. The root collar is part of the tree's trunk. Unlike roots, the trunk is not specialized to resist constant soil moisture. If the tree is planted too deep, or has soil or mulch mounded against the root collar and bark that is not accustomed to being wet, we will eventually see decline and ultimately failure of the tree at some point. The root collar from the start makes a difference. Strong root collars still experience a condition called wind throw, because support from the lateral roots may have been diminished, due to surrounding driveways, walks, streets or other factors, diminishing a stable lateral root system. We saw the entire tree, roots and all, lift out of the ground and topple over in the storm.

Nature demonstrates over and over the principle of decline and renewal. Whether it's governments, businesses or the oak tree in your backyard, decline and renewal is a double-edged sword, and is a natural order that opens up new pathways.

Change often comes from the bottom up, not the top down. Thomas Jefferson understood this principle as he commented on the Shay's rebellion of Massachusetts farmers in his often misquoted and misunderstood but famous writing. He understood that lethargy and ignorance on the part of the people would lead to decline when he said, "The tree of liberty must be refreshed from time to time with the blood of patriots and tyrants. It is its natural manure." I would like to think that he wasn't so much endorsing taking up arms to solve issues or causes, as he was endorsing a bottom-up, grass roots movement of knowledge and revival as being that natural manure.

Most people don't think of pruning or understand it. "Pruning" can be a good thing in both our personal lives as well as for a tree. A setback can help us re-evaluate and reset to move forward. A tree, even a large established tree, benefits from pruning. Often, trees that fall over in storms have been neglected for years.

Pruning of trees is a safety issue, a tree-health issue, and, finally, is done for aesthetics. Winter is a great time for pruning of deciduous trees and strengthens and improves them for future growth.

Lessons can be learned from the storms of life. Many times, the best people are those who have had their rear-end dragged backwards through a knot hole. With people, like trees, adversity can reveal true character and resolve. It can provide a well-rooted foundation for future storms.

I have times in my life where I feel like I'm buried, with a feeling of suffocation, and a feeling of frustration. I have learned that sometimes, when you think you're being "buried," you're really being "grounded," planted, in a manner of speaking, for *a season to grow.* My buried feeling was much self-imposed, thought instead of recognition and prioritization of what was ahead of me. The duress would be a springboard for *a season to grow.*

> 1. The tough part is "starting"
> 2. The tough part is "finishing"

Any initiative you undertake that causes some stress or has a fear factor, generally has two identifiable characteristics. Whether it is going back to school, starting a diet or exercise regimen like running, a new job, something you've never done, or, yes, even planting a garden and wishing for success.

1. The tough part is "starting"

2. The tough part is "finishing"

I have rappelled off a tall building, zip-lined from tall tree to tall tree in a forest, and run long distance races. I have observed, in each case, the tough part for people is starting, and the tough part is finishing. When zip-lining off a platform, affixed high in a large tree to a point you can't see hundreds of feet away, the tough part is starting. You have to jump off. Your approach to the next tree can be adventurous, as you either come up short and lose momentum, or overshoot your target and crash. It was exhilarating, jumping off tall *Liriodendron tulipifera* tulip poplar trees, hopscotching from tree to tree on a zip line. When zip-lining, I seem to have the anticipation part of landing down pretty good. Others in my group … not so much. Some people lose momentum and fall short; this causes frustration for the coaches who have to go out on the line to retrieve the fledgling airborne excursionist high above the ground and literally drag them back. Timing is everything. The tough part is finishing.

When I rappelled off a tall building, the tough part was extending a leg over the parapet with the ensuing tingly sensation, and then the second leg, putting the weight of my trust on the rope. The tough part was starting. Rappelling down the glass tower, I developed a rhythm that gave me comfort and confidence. Just as I was feeling good about myself I found that finishing was the tough part. The rope, now long and extended, did not provide the control I had previously experienced high above when starting. Now I know why you wear a helmet. The helmet isn't going to help if you fall. The helmet was there for the twisting in the breeze, and for the length of rope above that now allowed me to move along the face. When pushing off, if I would twist, as opposed to having feet planted, I had a sensation of rotation. My back was literally against the wall. I wanted control but there was too much "history" (as in rope) above me. I rappelled slower on the lower portion of the building than the peak. The tough part was finishing.

When I first decided to run long-distance, the tough part was starting. The distance was daunting, and you had to push self-doubt out of your mind. Once running, I was fine and settled into a rhythm. The tough part was finishing. Inevitably, a few miles from the finish, self-doubt would try to creep in. That is where mentally I taught myself to engage the Navy Seal rule. When your mind is telling you that you're done, you are only 40 percent done.

I had a chance to personally chat alone with Robert O' Neill, a former U.S. Navy Seal, special warfare operator and author of *The Operator.* He is best known for claiming to have fired the head shots that killed Osama bin Laden in the raid on his Abbottabad compound in 2011. We were backstage in Chicago on Navy Pier after his speech, talking about finding time and the discipline to run in our busy schedules. Talking about the required edge, Robert, in the true spirit of a Navy Seal said, "Complacency kills and success causes complacency." That's true, if everything was easy, and just came to us we would quickly become complacent. I'm not raiding compounds, so complacency might not kill me, but it certainly isn't going to grow me.

I love stories like the survivor Callery pear tree known as the "survivor tree," enduring the terror attacks at the World Trade Center in September of 2001. I was moved standing next to it and thought of all those who lost their lives on that fateful day. How those who remain rally, so that we never forget their sacrifice; it stands as a symbol of resolve, character and freedom. Or the 1,000-year-old oak (known as the "Big Tree") in Goose Island State park that survived hurricane Harvey, and still stands as a symbol of resolve and strength to Texans who lost so much in that tragic storm. Conversely, I remember a straight-line wind event that went through a community near me, where many large, stately trees failed in the storm. These trees were stately, tall and majestic pillars of the community. At a glance they appeared to be massively strong, immutable, enduring symbols that stand the test of time. Those that fell in the storm revealed they were hollow in their center. Upon inspection I will never forget the image of massive trunks on their side hollow to the core. We didn't know, it took a storm to reveal it. Their core was in decline, and all it took was one tempest to topple their position. I learned that, with both trees and people, adversity doesn't create character; it reveals it and exposes your core. May the forest be with you.

He didn't break any laws
With what the neighbors considered faux pas
The big takeaway
Was like a holiday
It made him feel like Santa Claus

Chapter Eight

Here comes Planta Claus

YOU DO A LOT OF MARITAL COUNSELING IN the work that I do. I call it foliaceous fighting between a couple, usually involving a pruners. Recently, I read a survey of the 50 things couples argue about most. Of course, money and kids were first and second on the list. An obvious issue like directions was high on the list, I believe number seven. What surprised me about this top 50 list is that gardening came in at number 11. They're feuding over the forsythias, quarreling over the quince, a bandy over the begonias. Yes, couples argue about gardening more than they squabble over the toilet seat, vacations or the in-laws. They'll bicker over their buddleias before quibbling over how long Uncle Harold and Aunt Edna are going to stay for the holidays.

Over the years we've had to settle an argument many times, being given the final say on their dispute. Often on phone calls to the show, you hear the spouse in the background, coaching the caller on what to say. We apply botanical therapy and don't charge for the session. I should, because over the years, I have become adept at mediating these horticultural marital disputes.

Nothing tops the legend of Malcolm Applegate, a British

man who hid in the woods for ten years to escape his wife's nagging about the time he spent in his garden. The more he worked on his garden, the angrier his wife became. She demanded he cut back on his time in the garden. Without a word he packed up and left. He left on a bike for London, traveling part of the way on foot, because his bike was stolen. He camped in the woods and worked on gardens at a community center for the elderly. He eventually moved into a homeless shelter in south London, where he worked in community gardens and raised money for homeless people in his spare time. This went on for ten years until, finally, his sister, who assumed he was dead, reconnected on the "telly" restoring their relationship. There's no word about Malcolm's wife. Malcolm has become legendary in gardening circles as an enterprising entre-manure.

I have a friend who owns a landscape and garden center, and tells me the toughest sale of a new landscape is to a retired couple. When bidding the job, decisions need to be made by the client. Picture the scene as he presents the plan to a couple standing in their yard. The man was a decision-maker away at work for 40 years, while his wife managed the aesthetics of the landscape. For 40 years his wife made the tough calls on where the rhododendrons would be planted and when the geraniums would be fertilized. No longer employed outside the house and with time on his hands, he's going to now make the biggest decision he's made today. He's going to step in and make the calls telling the landscaper how it's going to be. Meanwhile, his wife is standing behind him, shaking her head and making hand gestures. My friend struggles to keep a straight face, while the man's wife is giving thumbs up, thumbs down and rolling her eyes. She stands behind him waving her arms, shaking her head and making hand gestures, mouthing silently her disagreement. Selling a landscape job to a retired couple is a delicate diplomatic dance. Having managed the flora of their household investment for 40 years, she's grounded in what will and won't work. It requires diplomatic sensitivities on the part of the landscaper to close this deal before a shovel is put in the ground.

I remember a phone call from John, live on air, who wanted to prune his Japanese maple. Typical of Japanese maples, an early May frost, after they have foliated, will nip the leaves at the top of the tree. The damage is unsightly only for a few months as the tree naturally re-foliates as the season progresses. The conversation on air goes something like this.

"The tree is too big anyhow so I think I'm going to prune it."

"Absolutely, go for it John, pruning isn't going to kill the tree, it will cause fuller growth."

"Really?"

"Yes, if you don't want to wait for new foliage to grow out and the tree is too big, have at it and cut away, but wait of course until the end of the show."

"I better wait until my wife comes home."

"You can do this John."

"You're right. I can do this. After my wife gets home. I don't want her pulling in the driveway seeing what I have done."

John called me back on the show ten months later. He said it was Easter weekend, and he was hoping his wife would now finally forgive him for pruning the tree. It occurred to me pruning is inevitable in all our lives. What we all need is redemption and forgiveness. Redemption and forgiveness is the ultimate beautiful gift afforded us all.

There is a big old oak tree along a path where I walk to the beach. Weathered and worn, yet majestic and stately, this tree has seen its share of seasons, and experienced its share of storms. Renewed in spring and fallen in Autumn, if only that tree could talk … oh the stories it could tell. Walking alone I have conversations with that tree. Well, I do all the talking. That big, old wise oak is a good listener. The unrelenting strong winds off the lake have pruned this tree through the years. It stands firmly rooted, the years of time etched on its canopy. The trunk is massive and furrowed. If a tree can be wise I imagine this one to be. When I gaze on the tree I think of the phrase, "Everybody gets older but not everybody gets elder." I like to think this old oak got elder.

A shade garden is inspiration from the wild. Who doesn't love a walk in the woods? There is something mystical and magical about a woodland community. Trees in a woodland area are not obstacles. They shorten the views giving the space an intimate feel, while providing a roof for the understory activity. Trees both dwarf the person and the views. They provide a stillness in space that breeds discovery. That is why a path is so effective in garden settings. It is fun to imagine and then see what is around the next corner. Learn to establish focal points that draw you down a path; it's enjoyable, because when you reach that point … you've arrived. You can dwell in the moment for a while and enjoy its beauty.

Shinrin-yoku is Japanese for forest bathing. No, it's not taking off your clothes in the woods and taking a bath. It is immersing oneself in nature, taking in the forest atmosphere as a gentle path to wellness. Considered by some to be trendy, I think it's a great healing practice rooted in culture and history. A practice of taking in the forest atmosphere and breathing as a form of medicine. Scientific studies are now proving what we have always intuitively known, that nature can reduce stress levels in humans. Moods are improved, and healing is accelerated.

This is part of the reason foliage plants are encouraged for indoor "breathing rooms," or, as I like to call them, "jungalows." These are not just trendy concepts. Unplugging from our electronic devices and opening the senses in a forest setting or an indoor breathing room, might be what you need to lower your blood pressure. Maybe a chance to be idle for a while and do nothing to enhance your creative juices. The Dutch have a concept of Niksen, derived from the Dutch word "niks," which can be loosely translated to mean nothing or do nothing and be idle. My favorite word in this regard comes from Norway, with an ethos all its own. Friluftsliv, literally means "free air life" in Norwegian. A spirit of friluftsliv is good for what ails us, whatever the culture or country may be.

> Phytoncides, the essential oil produced by plants, has been credited with lowering stress and cortisol levels.

We now know it is more than the peace of the forest, the unplugged feeling of fresh air and the exercise of a quiet hike through the trees. We now know that phytoncides, the essential oil produced by plants, has been credited with lowering stress and cortisol levels. Aromatic volatile oils are produced by trees, like pine and hemlock, as protective agents, and are credited with the ability to lower blood pressure. The phytoncides, released by trees and plants into the surrounding atmosphere, is done to protect themselves from disease and harmful organisms. Inhaled phytoncides from a walk in the woods can last days in the human body. This goes way beyond the scented pine tree, hanging from your rearview mirror, masking the odd odors in your car. These natural oils are bathing you in stress reduction, resulting in a healthier you. It's the gift that keeps giving.

REJUVENATION PRUNING OR "STUMPING" OF AN OLD OVER-grown landscape plant like a taxus, lilac or rhododendron, can give new life to the plant and change everything. It's convincing the owner that everything is going to be alright. If the owners are a couple, it gets tricky, because they begin to question each other. The plant is going to look ugly for a while until it rejuvenates. That's when doubt creeps in, while you're living in the unseemly stage of the process. Actually, it is a metaphor on life as we know it. A setback initially looks bad, but as my friend Stephen reminds me, as with anything, "Rick, it's always not as good as it seems and not always as bad as it seems."

It's best not to take a laissez-faire approach to your French lilacs. Cutting them back right after they are done blooming will enhance growth, sunlight penetration and future flowering. Managing their growth can be an investment in future blooming. Pruning is a good thing. Odds are the deer aren't going to prune your lilacs for you. Lilacs are considered "deer resistant" in gardening circles. Yeah, right, tell that to hungry Bambi in February with thick green buds on the ends of the stems. Again, the old adage applies, if a deer is hungry enough, they will eat just about anything.

Money does grow on trees. Leaves are an incredible resource falling at your feet. In our suburban neighborhoods, we manicure the lawns and gardens, so, in October and November, every leaf is sought out and blown, bagged, tagged, raked and disposed. I have a neighbor who feuds with a neighbor two houses down. He waits until they leave their garage door open on a windy day, and then conveniently blows his leaves off his lawn in their direction. The neighbor finds their garage full of leaves. In fall we end up moving leaves around and sharing with the neighbors. A blustery day distributes fallen leaves with neighbors better than a porch-lit trick-or-treat candy distribution on Halloween night. Everybody wins!

If there is nutrient value in the leaves falling to the ground, why don't we celebrate it? I made the mistake on one radio show to mention that "I was in need of leaves." The donations poured in. A windfall of benevolent goodness, it reached the point where I had to hide, so that well-meaning philanthropists would stop bequeathing their bounty. I needed a different strategy.

I hatched a plan that would utilize the onset of daylight saving time in early November. If I left work at 6 pm, under the cover of darkness, I would pick up bagged leaves people left at the curbside. I looked like "Planta Claus" carrying his big bags of toys. Operation Planta Claus was born. I would pick them up under the cover of darkness with my truck and dash for home to recycle them. It was like Christmas in fall and my composting skills were "elf" taught.

> *As dry leaves that before the hurricane rustle*
>
> *He would park at the curb and the bags he would hustle*
>
> *Into the bed the leaf bags flew*
>
> *With a sleigh full of leaves and Planta Claus too.*
>
> *He would pack up his ride with foliage to haul*
>
> *Now dash away dash away dash away all!*

As part of my horticultural high jinks, I would scope out the neighborhoods that had lots of big maple trees, and plot my raid for after dark. They did the work raking and bagging, and it was mine for the taking. Once those leaves moved to curbside, it was public domain, and to the swift belong the spoils. Considering it foliole folly, they didn't understand that it was supplementing my income. Not that I have to use a form 1040EZ, I recognized this bonus stipule would reap rewards down the road as a tax-free windfall.

Here comes Planta Claus, here comes Planta Claus

Right down Planta Claus lane,

So bag your leaves and say a prayer

Cause Planta Claus comes tonight

I had based my assumptions on a Rutgers University study, where I had read that 100 municipal leaf samples from across New Jersey were collected to be analyzed as a source of crop nutrients. Obviously, nutrient concentration values varied considerably, but there was definitely the big three: Nitrogen, Phosphorus and Potassium content in the foliage. In addition, essential micronutrients such as magnesium, calcium, iron, manganese, copper and zinc were also present.

Here is where it gets really cool. The foliage would, as it does in the woods, contribute to the long-term fertility of my plot as the nutrients are released over time. The organic structure of the foliage requires microbial decomposition to release them. The benefit of organic matter, improving the tilth and structure of the soil, as well as increasing earthworm residence, was a bonus benefit. The nutrient value would be enhanced. Of the three major nutrients, the most readily available in the first year from leaf waste was potassium. Potassium is an essential plant nutrient for proper growth, cell-wall development and reproduction of plants. It also is a major contributor to the overall health and vigor of a plant. That's why my Mom always wanted me to eat bananas, beets, potatoes and beans! Protein and starch synthesis in plants requires potassium. It also plays a big role in how

> When I find new plants at the garden center, I walk around the yard with a shovel, looking for a place to put them.

a plant "breathes" with the opening and closing of stomata in the foliage regulating CO2 uptake. Both the uptake of water, through plant roots, and the loss of moisture, through stomata, are affected by the potassium levels in a plant.

I took it to the next level with what I call "lasagna passive composting flower bed construction." Lasagna in that the soil amendments are placed in layers; passive, because it doesn't involve much physical work. I was always converting lawn area into flower beds. I find some more new plants at the garden center, and am next walking around the yard with a shovel in one hand and plants in the other, looking for a place to put them. That is why every fall I would create new planting areas, reducing the turf space into planting space.

It's easy and effective following this procedure. Mark out with spray paint the size and shape of the bed you want to create in the existing lawn area. Instead of paint, you can also use a garden hose, and move it around to view and reshape the border. In mid to late October, as the leaves are coming off the trees, lay newspaper three or four pages thick over the top of the grassy area within the assigned border. Don't try this on a windy day, or your neighbors will be picking up trash for days. Once the paper is laid flat on the grass, wet the paper with a garden hose to hold it in place. Next, place leaves fallen from the trees on top of the newspaper. I prefer to run over the leaves with a lawn mower once or twice before placing them on the paper to speed decomposition. Once the leaves are liberally put in place within the boundaries of your new bed, distribute soil and compost on top of the leaves to hold them in place for winter. The snow and rain can now begin to fall. Over the course of the winter, the newsprint will suffocate the grass below it, so you don't have to dig it up or use a sod cutter. In spring you can till the leaves, newspaper and top-dressed soil into the parent soil and dead turf to create a rich and ready flower bed for spring planting.

Shade gardens are best when enjoyed through filtered shade. Structural shade is unforgiving and dark. Filtered shade caused by trees is seasonal and variable. Light is rewarding and dances in the space. Dark spaces needed to be lit with golden foliage or variegated foliage.

Sun plants tend to have smaller leaves, waxy coatings, hairy foliage or a powdery look, all designed to help them conserve moisture. Shade plant leaves tend to be bigger, broader and thinner. Smooth and shiny, they are able to wash off dust and grime in a rain shower, maximizing their leaf surface for available light absorption. The larger the leaf surface, the better it can capture available light and shade the roots of the plant. The foliage becomes its own living mulch, able to shade soil temperatures and shade out weeds that compete for moisture, space and nutrients. Overlapping foliage, which is common for shade plants like hosta, creates amazing landscape textures and density to allow it to compete.

I watch the *Ficus benjamina* trees that are shipped from Florida to Michigan. When they arrive, their glossy leaves have a very pronounced midrib. That's the visible line right down the middle of the leaf from petiole to the tip. Over time, as they endure the cloudy, low-light seasons of Michigan weather, that midrib begins to disappear on the foliage. Important, when it lived in the Florida sun, the midrib would cause the foliage to fold up in half to conserve moisture. Not needed in Michigan, the foliage indoors becomes flat, trying to absorb what light it can. In addition, seasonally up to half the foliage drops off the plant as a defensive mechanism. When that happens, if you run for the watering can and apply more water, you simply rot the roots and speed the decline.

Here I see a correlation between the landscape and our lives. When plants are situated in a shaded or dark place they adjust. They adapt to the environment they dwell in. If the environment is wrong, they either curl up and die or adapt and thrive. The metaphor on life for me is "you can't change some realities of life, but you can change how you experience it." If the sun isn't shining, make your own sunshine, and things will get better. Right plant, right place. When plants are respected, by being properly located, they thrive and find relevance.

Consider with me the snow, when it falls on the trees, making a panoramic, awe-inspiring sight the morning after a snow storm. It's beautiful to see, but what is the burden to bear for our shady tree canopy? A rough estimate of calculation I use to determine snow weight is 1 inch of snow = 1 pound per square foot. This is variable based on the outdoor air temperature. Late fall snow or early spring snow tends to be slushy and wet with a high-water content. If it's very cold outside the water content could be five percent of the flake. If the temperature is warm enough to be around the freezing mark, then we get the wet, heavy snowflakes that could be as much as 32 percent water content. The cold wind-blown five percent water content flakes would be around .26 pound per square foot. The wet, slushy, late fall snowflakes could be 32 percent water content or 1.66 pounds per square foot. So, the generally accepted rule of thumb is this: figure around one pound per square foot. I have been on my share of roofs and greenhouses through the years, shoveling snow loads, and, by the end of the day, can attest to the backbreaking weight potential of those innocent-enough looking snowflakes.

> You can't change some realities of life, but you can change how you experience it.

In October and early November, snowfall can be devastating to trees and shrubs, because the water-content weight tends to be heavier, and if the plant hasn't dropped its leaves yet, the surface area for it to collect more weight is significant.

I love to look at a tree and try to guess how many leaves are on the tree. Here again I apply some unscientific simple math to target a number. Start by pacing out the length of the canopy of the tree in an east-to-west direction. Next, pace out the distance of the canopy from edge to edge in a north-to-south direction. Let's say it was 50 feet in each direction. 50 ft X 50 ft = 2,500 square feet of tree canopy. I multiply by 4 to get a total of 10,000 square feet of leaf area. Then, laying some leaves on the ground in a one-square-foot area, for example, let's say you fit eight leaves in the box. Ten-thousand square feet of leaf area times eight leaves per square foot equals 80,000 leaves on that tree. Now I'm sure some math wizards can poke holes in my method, but it works for me, and, if you don't believe me, you count them.

If in fact, there are 80,000 leaves on that tree, it brings up some interesting things to consider. Its thinking about these things that gives me an appreciation and wonder for the trees around us. First, we know leaves produce oxygen and transpire, which cools the air. The loss of moisture in the foliage creates an amazing "pull" or pressure in a tree's built-in plumbing system to bring water from the roots. Now, I hate plumbing, because no matter how big or small the job, when I'm done it always leaks. After four or five trips to the hardware store, I would have been much further ahead by calling a plumber in the first place. Secondly, I am amazed how leaves produce food. Plants feature a characteristic called autotroph, which is their own ability to make food. Unlike animals, who can't make their own food, plants consume the sun and transform the energy into starches and sugar through photosynthesis. I think of this every time I watch the hungry deer from my window, eating my roses and hydrangeas. Look at a tree and imagine the branches as arms and leaves, as hands extended palms up to the heavens to make their own food. This makes a plant, in many ways, more resourceful than an animal. This is the reason millions of dollars are spent on animal repellents for our landscapes every year. Many humans can't even prepare their own food, and are dependents to the drive-thru window. Ahh, the amazing ability of plants to make their own food. I read this quote and it really resonated with me:

> *The most important chemical reaction on earth is photosynthesis. We are all parasites upon it.*
> *– Robert DeFeo National Park Service*

The really cool part is this: the leaves fall to earth in fall as a natural resource, that is, nutrients and organic material. Once again, the tree is feeding itself. In the suburbs, where we cut down the trees and then name the streets after them, we bag, tag, mow, blow, rake and dispose of this natural resource. I am a big proponent of a quality mulching leaf blade on a mower to finely mow the leaves back into the turf and garden. It is estimated, over the course of its life, a shade tree will shed around 3,000 pounds of leaves, returning nutrients back to the soil.

So, back to my mathematics lesson. (No matter how flawed or simplistic it may be to my 4.0 GPA calculus friends out there.) If I am correct that the tree with a 2,500 square-foot canopy has 10,000 square feet of leaf surface area, with eight leaves per square foot or 80,000 leaves, and a wet snow in fall weighs 1.66 pounds per square foot, the weight of a three-inch October snowfall on that tree could be as much as 49,800 pounds. (1.66 pounds per square foot, times three-inch snowfall, times 10,000 square feet. That's really heavy man … a burden on my mind.

After a storm of heavy snow, we see the weight bend plants to the ground. If the evergreens or trees do not break under this extreme weight, they will bend. If left bent for a prolonged period, it can change the cell structure in the branch or branches, causing permanent damage, not allowing it to bounce back. That's what living in fear does to us as people. The weight of negative circumstances or a negative environment bends us. We are affected by its influence and can internally change our approach to life and our contribution. Positive people are like sunshine, and a negative environment can have lasting circumstances. Surround yourself with positive people.

YOU'VE GOT IT MADE IN THE SHADE

WHEN MANAGING A SHADY AREA HERE IS MY TOP TEN LIST OF things to do

1. Don't lose your composture. Some of the most beautiful gardens are in shady areas. Don't throw in the trowel just because you have shade or root competition from trees.

2. Understand the type of shade (seasonal, structural, filtered)

3. Use shade-tolerant plants. Foliage is usually larger on shade plants to absorb available light and shade the roots.

4. Amend the soil with lots of organic material. (Think woodland floor.) Soil should have moisture and nutrient-retention capability, but be well drained with drainage capability. For example, rhododendrons can get quite large but have small root systems, proportionately. If they sit in water, the plant will suffer and die.

5. Don't overwater low-light plants.

6. Modify shade conditions with winter pruning. Take an in-season picture, and then get out there in winter to do some maintenance pruning.

7. Account for root competition from trees. Tree roots do not go to China and compete for moisture and nutrients at the surface. Again, add lots of organic matter, thinking woodland floor.

8. Use groundcovers or mulch, where possible, surrounding the trees. Turf and trees are not compatible, not just because of shade, but also because of root competition.

9. Use containers in shade gardens. Containers can be moved to improve light conditions. The containers provide interest, color and character in the shade. Also, stone and natural concrete statues weather well with the occasional lichens and moss.

10. See #1 above. There are many great plants to grow in the shade.

About to give up on Zucchini
With no bottle, no wishes no genie,
She expressed agitation
I said try artificial pollination!
Look under the bloom for a bikini

Chapter Nine

Lettuce Party like it's 1999

'M FEELING VERY VERDURE ... A CONDI-
tion of health and vigor associated with green, growing
vegetation. Imagine that at my age vegetables could
make you vogue. Know your food, grow your own, local urban
farming and vegans have made vegetables voguish. What kind
of food do you want to be associated with? An important question, because we take our food and food choices personally.
Healthy soil, healthy plants, healthy you. Healthy soil, sunshine
and irrigation practices are the primary keys to vegetable nir-
vana. Practice this and you will look verdure to your friends
and family.

In the language of flowers and plants, a red rose means love,
and a marigold means guilt. A potato, however, means benevo-
lence. If someone gives you a potato, they are just trying to be
kind and well meaning. Remember, some people are like potatoes; they might be thick-skinned, but are soft and warm inside
when warmed.

Growing up in the '60s and '70s, I did have a green leisure
suit and silk shirt. We were stayin' alive and dancing to the Bee
Gees. Today I'm not a vegan, but I have made it to flexitarian
status, where you can go a day or two without meat. I also try to

fast a day or two each week with reduced calories. It strengthens both my mind and my body and keeps me young. I am all for intermittent fasting and a flexitarian diet, with exercise and a dash of chia seeds, apple cider vinegar, turmeric and I'm good to go.

> Back in the '70s as a teen I thought kohlrabi was a middle east country.

Back in the '70s as a teen I thought kohlrabi was a middle east country. Today, I drink beet juice, eat fermented vegetables, and put kale on my burger. Breakfast is cooked oat bran and blueberries. A word of caution: when drinking beet juice, don't forget to wipe your mouth. I have, and have gone out in public looking like the Joker with a grin akin to Jack Nicholson in the Batman movies. This journey has been "veg"-ucational, and started for me at a young age when I worked on the farm.

MUCK FARMING IS DONE ON DRAINED SWAMPS, AND IS where many vegetables are grown in Michigan and northern states. American "muckers" have their roots from the Netherlands or Eastern Europe, where our ancestors grew root crops like onions, carrots, parsnips, turnips and potatoes. My ancestors were from the Netherlands, so I was a proud mucker, carrying on the family tradition.

The muck was a problem on windy, dry days, because you would blow your nose and the tissue would be black as night. Your clothes were dirty as well. The washing machine got a workout and your work clothes never got fully clean. With black faces, nostrils and knees, we looked like coal miners and subject matter for a Johnny Cash song.

That black muck soil could get hot, real hot. It was burn-your-knees hot. There are stories of that soil, self-combusting and burning in summer. Don't get me wrong, I love hot weather, I just don't like hot muck. I am a hot and humid weather guy. I can't help it, that's when I feel most alive. Forgive me I am only "humid." I just "pun-ditioned"the air. I do not like air condition- ing and do not feel well when subjected to it. It is one thing to "condition" the air, it is another to create a meat locker situation, where a side of beef would be frozen solid. I do not like sitting in a vehicle or movie theatre, where the conditions feel like you're in Oymyakon, Russia. Having to wear a winter coat in August, while the nose hairs freeze in your nostrils, is completely ridicu- lous. If I wanted to live in Vostok, Antarctica I would move there. That is why, when it gets warm in the great North, I spend every possible minute outside soaking it in. Take your example from the plants, they thrive when it's warm and look naked and sad when the temperatures are near zero.

The dictionary definition of muck (as a noun) is *dirt, rub- bish or waste matter*. Synonymous with grime, filth, mud, slime, gunk, glop and grunge we spent our days in it. Like a grunge rock and roll band looking for trouble, we would trudge into the fields for our daily gig. Crawling on hands and knees, digging, pulling, planting and bundling, we would spend hours vegetat- ing. The dictionary definition of muck (as a verb) is *to mishandle a job or situation.* And we did that too as in, "boy I mucked up my Algebra test today." We were proud muckers … broke, dirty and living the dream!

During those long days or evenings in the field, we didn't have access to … let's call it restroom facilities. No rest for the wicked. A solitary, dilapidated outhouse was positioned near a ditch, and, as cheap as my boss was, it wouldn't surprise me if he rented out the basement. He was so cheap that toilet paper was not provided to the laborers. A phone book was provided for that purpose, and when we got to Zeeland township in the phone book, it was time to bring a new one in. I would always sneak into the outhouse when nature called. If anyone knew I was in there it was liable to be tipped over as I sat on the throne. Occasionally your partners in grime would hollow out a radish and press a Black Cat firecracker into the red orb, light it and throw it in the house. Fire in the hole! You would pick radish specks off your face and your ears would be ringing for the next hour.

Radishes are nasty projectiles … not as bad as rutabagas but effective when they find their mark. They leave a welt and were often used for muck-farm warfare. A large lumber outlet adjoined the farm property, and when a poor unsuspecting hi-lo driver would dare venture around back, he could be targeted with a barrage of vegetation. An herbage skirmish would ensue, and, if the hi-lo driver reported this to his boss at the lumber yard, there would be hell to pay. We could hear boss man fire up his motor scooter at the wash house, and soon a cloud of billowing dust was visible on the horizon, rapidly closing on our position. When he came into view it was a sight to behold. His face was as red as a radish, and his large physical frame was at least four to five times larger than the laboring motor scooter. He was so big, and the scooter so small in the dust storm it created, that he almost appeared to be levitating in an angry cloud of soot. Wearing a dirty, old baseball cap and an untucked flannel shirt flapping in the breeze; this was not Santa Claus coming to town. It was more like a mushrooming cloud of vegetable Armageddon. As he approached, all the workers quickly held a less-than-democratic forum and straw vote, as to who was going to be thrown under the bus when he arrived. The sacrificial mucker would be sent home for the day until the boss cooled off, instead of all of us having to stay late. I'll never forget his imposing size as he would dump the bike and approach, yelling at us hotter than a salted radish.

I remember kneeling with my box of onion sets at the start of the row as long as a football field and looking out over the horizon. If Phileas Fogg could circumvent the world in 80 days this could take as long or longer. Peering across the scablands that lie ahead in the scorching heat I would soon begin to see a mirage of cool drinks and swimming pools as I crawled along the hot soil. In the spirit of Hernan Cortez, you had burned the boats, because there was no looking back. Your conquest of the row was defined, and you weren't going home until it was done.

Worse than having to plant long rows of onion sets crawling on hands and knees in the dark, hot and dusty soil, was picking the green onions once they grew. When I had planted the onion sets weeks earlier the boss would pass by with a spreader, applying what I later learned was Diazinon. An organophosphate insecticide banned by the EPA to sell for residential use as of December 31, 2004. I can still taste the residue on my tongue and recall the aroma in the fields. I quickly learned that onion-root maggots would feed on the developing bulbs, encouraging the entry of soft-rot pathogens. As I pulled the onions to bunch them, I slid my hand along white ends to peel the outer skin, and quickly discovered that some were rotted and smelled. A delightful process, so by the time I got home, I smelled like Diazinon, rot and onions. My hands were always filthy no matter how much I soaped them. I eventually invested in Alka Seltzer and would soak my fingers in it to clean my nails. Plop plop fizz fizz oh what a relief it is.

The best part of the week was Saturday at noon when I got my paycheck. It was usually all of ten or twenty dollars. We would proceed en masse on our bicycles to the local grocer called Noel's. With a parking lot full of bikes outside, we would converge on the store and cash the fruits of our labors, buying lots of sugary junk food in the process. Sometimes a lucky few could ride in the back of the stake truck to market and get an ice cream cone on the way. You had to eat fast or the cone would melt in the wind.

Slowly but surely, after 1999, the year I turned 40 years old, I realized that, just like those spring planted plants, I was vulnerable and wasn't going to live forever. It wasn't so much the fact that my days were numbered as it was "what would the quality of life be for me in the second half of my life, when I possessed the wisdom I didn't have the first half of my life." Now, I had a chance to flower and produce better than I ever had before. I shuddered when I thought of how I had treated myself in the first half. It was time to party like it's 1999!

I didn't want to be depressed and out of shape during the second half of my life. It was at that point I decided that, even though a group of "old" people would die at a similar age, I was going to make the last years a thriving experience versus a surviving experience. If you were to create a curve on a graph you might find the majority of us die at about the same time in life. I get that, but would I rather the curve have me go down in flames in a plummet or a slow gradual decline? How do you feel my friend? If we both die at 90 do you want to experience a full life when sage and wise at 80 with the capacity to be a flowering member of society? I have learned a sage, healthy man at 80 is far more dangerous and effective on influencing the future than a man who is old and has given up on life. I have seen plenty examples of both. My doctor calls it "no dwindle" to the end. When I was younger I didn't understand people who said they wanted to come in for a landing at full speed and squealing tires but now I understand. Square off the curve. Not spending the later years re-"tired" but rather re-"wired" for quality of life. Our lives might end at the same age, but the quality of the later years, and how I feel, is something I can invest in now. I know what to do, thank you very mulch.

> My doctor calls it "no dwindle" to the end. Square off the curve.

We have a love-hate relationship with a number of things in our life. Exercising is one. We pay someone to cut our grass for us, and, in turn, then pay the health club so we can run on one of their treadmills. Zucchini is another. We buy plants and amendments and spend hours nurturing the plants only to force them on others in August. There is a day designated for national share zucchini with your neighbor. Put them in a grocery bag on their front porch, ring the doorbell and run.

People love vegetables, but many are polarizing with a vegetative love-hate relationship. Maybe that's why we love heavy dressing. Cilantro or coriander tastes fresh and citrus-like to me. There is, however, a large group of people who feel cilantro tastes like soap or metal. Kale makes you feel good about what you're eating, but I've had people tell me it tastes like ammonia. Working in retail throughout the years I've observed that same frustration with the Christmas season … the only time of the year we bring a dead tree in the house and fill socks with candy. You get a day off and spend it with some relatives you try to avoid the rest of the year.

Now when I say love-hate, I mean loath more than I mean loathe. Loath is a description of unwillingness. Loathe is an action to hate intensely. Hate is a strong word. But to loath a vegetable means "when served it is soon destined for the compost pile."

NOTHING LIGHTS UP THE PHONE LINES ON MY RADIO SHOW quite like a conversation about food. Debate over food is a polarizing topic, and taste is a personal subject. People take it personally using varied adjectives to describe their position. Many vegetables can be classified as classic like-dislike polarizing subjects, from brussel sprouts to beets, celery to cilantro, olives to mushrooms and kale to lima beans, everyone seems to have an opinion. Arguments break out on how to prepare your brussel sprouts. For me, brussel sprouts are best roasted in the oven, sliced in halves and then slathered in maple syrup. Others want bacon with their brussel sprouts and still others would rather just pass altogether. Are lima beans a rich and buttery delicacy, or do they taste more like wallpaper paste? Some will defend lima beans voraciously and attempt to make converts of others.

How about green bean casserole? That all-American dish synonymous with Thanksgiving and holidays. I am thankful for the nice lady, versed in home economics, who created the recipe in the test kitchen of the Campbell Soup company. A can of Campbell's cream of mushroom soup, some beans, milk, spices and dried fried onions on top is one big bowl of delicious genius. There seem to never be enough of those tasty dried onions floating on the top. At least that's my opinion, because even if you want to argue with me you can't argue. That my friend ... is real comfort food.

I love cucumbers, especially cucumber salad. A caller suggested I eat them as "cucumber boats" slicing the cucumber lengthwise down the middle and hollowing out a "boat" filled with peanut butter.

Of course, okra is another polarizing produce that seems to cause people to simply ask ... why? I tell them I like it because I'm from Okra-homa.

I love the tradition in the Japanese town of Minoh City, where they deep fry the maple tree leaves. It's not fast food. It takes patience and about a year to prepare. Perfect yellow leaves from maple trees are hand-selected, and then packed in salt and left for a year. The salt is then shaken off, and the leaves are individually deep fried in tempura butter coating, with a little bit of sugar for a truly regional delicacy. "Maple" I will try this when less pressed for time and retired. It is a taste treat that could catch on.

We won a Michigan Association of Broadcasters award for a segment we did called *Poke Salad Annie*. Call it polk or poke and call it salad or sallet. Poke sallet, is the cooked greens-like dish made from pokeweed, and much revered in the 1969 hit song *Polk Salad Annie*, written and first performed by Tony Joe White and made famous by Elvis Presley. No one could or ever will be as cool as Elvis in a white jumpsuit on a Las Vegas stage singing about an awful tasting salad.

Regardless of what cultural background or traditions you ascribe to, this is one salad that has got to taste terrible. Made from the leaves of pokeweed, which tends to grow in unkempt areas like creeksides or roadsides, I would take a pass as all parts of the plant leaves, stem, berries and especially the parsnip-like thick tap root are poisonous. It might be a great song, but I would pass on this southern delicacy. Now celebrated in some annual festivals, people ate it because they were poor, and the plant itself is similar to kudzu, a bane in the landscape growing so fast it requires a new zip code by the end of summer. Birds drunkenly feast on the dark, purple berries at their own peril, risking a beak-stand on a clear glass window.

We all know that we need to eat our vegetables. Many people need more vegetable servings in their diet. A great part of gardening is the satisfaction of growing your own food. Today, more and more people want to know where their food came from. I call tomato plants and pepper plants the gateway drug to gardening. You can convince most people to try to grow one, and, if they are successful, they are hooked on gardening, expanding the choices and quantity they plant in subsequent years.

Tomatoes come from tomato plants, and, I know eggs come primarily, but not exclusively, from chickens. We have interviewed experts who feel duck eggs are far superior to chicken eggs. I have nothing against eggs. I enjoy a good hearty breakfast of vegetables like spinach and peppers diced and mixed with my scrambled eggs and a piece of toast. I, however, publicly made the mistake of admitting I had never had a deviled egg in my life. This resulted in a furor as all the deviled egg lovers came out in force to question my sanity. How is it possible a man in his fifties has never had a deviled egg? Maybe because they smell disgusting and the texture is gross?

I decided to do an impromptu survey in social media. The response was overwhelming. About 90 percent of respondents absolutely loved deviled eggs, and ten percent were in my camp of dissension. Overwhelmingly, the response was "I needed to live life a little." Take a walk on the wild-buffet side.

I said, "There's a reason they put the smell of these rotten eggs in utility gas lines so we can tell if there's a leak." People tried to convince me it's like an egg salad sandwich without the bread. They started arguing whether you should use mayonnaise or Miracle Whip. This really stirred things up. People started talking about food poisoning and ending up in the "Mayo" clinic. A common phrase was "I like them but they don't like me." A nice way of saying they stink and if they smell bad going in odds are they're going to smell bad going out as a gas. I was encouraged to plug my nose or eat them whole when eating them. That's a warning sign right there. I was told not to be so hard-boiled and to come out of my shell. I wasn't about to "whisk" it and might be shell-shocked if I tried. People talked of adding mustard, paprika, jalapeños, sweet pickle relish, dill pickle relish, pepper, horseradish, onions, bacon, olives, thousand island dressing, everything but the kitchen sink. Which is where they belonged, in the sink disposal.

> *A guy walks in a bar with a fried egg on his head.*
>
> *Bartender says, "why do you have a fried egg on your head?"*
>
> *Guy says, "because deviled eggs roll off."*

Well finally the challenge came to eat one live on air. The gauntlet had been dropped and I wasn't about to "crack." Adding insult to injury the number of postings in social media "egg-sag-erated" the challenge, so that I would have to perform on both radio and live social media on camera. Chef Mick would make the eggs and deliver them to the studio.

Thousands watched and listened as I forced half of the malodorous cuisine in my mouth and bit it off. A sulphurous burst, with a rubbery texture, it was all I could do to gag half of it down.

Fortunately, Chef Mick had brought along sufficient Bailey's Irish Creme which I kicked back to aid in the descent to my stomach. After regaining my "composture" I proceeded to send the remaining half down the hatch with ample gulps of spirits. Numerous people since that day have told me, "you should have tried mine, they're the first thing gone when I bring them to parties, even before the fudge!" Right, I would throw them in the trash and eat the fudge. One nice lady tried to convince me it was a visual thing. "If you're not going to close your eyes then fancy them up!" she said. She suggested I spiralize zucchini noodles and dress up the offending comestible. I told her a zucchini is just like a human, 80 percent water without the anxiety.

It is a fact, whether you like it or not, that zucchini is tough to give away in August. Fill a grocery bag and place it on your neighbor's front step. Ring the doorbell and run. Hopefully they're inclined to make some bread. It is quite telling that we make bread from zucchini. It's like bananas. When all else fails make bread out of it. When we make bread out of something, it's possible we may have given up just a little, because man shall not live by bread alone.

Barbara called me one day. She insisted the zucchini plants she was sold by a greenhouse were men. Frustrated by the lack of pollination she was ready to give up. She had called the greenhouse where she bought them and was told, "You have males but don't worry they will become a woman." Barbara exclaimed, "That's ridiculous. I went on the internet to figure it out and I guess I just have men." How could we produce baby zucchinis when she had two male plants? Barbara surmised she got two males because the plants were cheap. (She had spent only a buck.) I said a buck for a whole zucchini plant was pretty good, because corn was a lot like pirates, a "buck an ear." Barbara gave me a courtesy laugh. Before forming an opinion of men and heading to the farmers market I suggested she try the Q-tip approach. Giving her permission to lift the blooms and take a peek under there, I wanted to help her find a female live on the air. If Barbara would get up close and personal with some of the blooms, she would find some with miniature zucchinis attached to the base of the flower; then she would have then found the women. She took the phone outside and sure enough to her amazement we found a couple of females live on the air. I recommended she find a Q-tip and transfer pollen from the male flowers to the female flowers, giving the birds and the bees (and the wind) a helping hand. Barbara's reaction was she "still doesn't like kale." She did cast a vote for lima beans however. I loved Barbara's sense of "humus" and we encouraged her to try again and not give up on zucchini. Barbara loves stir-fried zucchini making a meal of it.

A colorful way of living, more of us are adding purple to our diets like eggplant, beets, purple potatoes and blueberries. We dream of the past when fast food and drive-up windows were not an everyday occurrence. We long to eat outdoor food outdoors, instead of something in a paper container that messes up our car. Know your food and grow your own. If someone decides to eat healthier and know their food, they often start with tomatoes and peppers. Tomatoes and peppers are the gateway drug to gardening. Have success with them, and soon you'll be hooked trying to grow everything from cucumbers to cilantro.

When trying a few tomatoes remember to look at the tag of the tomato plant you are buying. You will see somewhere on the tag, it will tell you if the plant is indeterminate, determinate or semi determinate. This has nothing to do with the plants determination to succeed. An indeterminate tomato will grow large and not set a terminal bud, meaning that if it didn't freeze come October it would keep growing to the size of a very large shrub! These are best planted in the ground with good plant supports to hold them up and allow sunlight and air movement around the plants. A determinate tomato, however, does set terminal buds, so they stay at a more manageable size, making them perfect for container growing. A semi-determinate tomato is obviously somewhere between the two in size, needing support but can be grown in large containers or in the ground. With any container gardening, remember to make sure that plenty of room is available in the pot for root growth and stability. They're going to grow!

To grow great tomatoes, make sure you have a good sunny spot and have prepared the soil with lots of organic matter. Tomatoes like a moist, well-drained soil, and, if the soil gets dry between watering, you are likely to have cracking and zippering of the fruit. They also like a boost of calcium to avoid blossom end rot on the fruit, so look for a fertilizer that has micro nutrients, including calcium. I like to use a complete fertilizer with major and minor nutrients, working it into the soil at the time of planting, and then top dress again during the growing season.

Another way to ensure good moisture availability is to deep plant your tomatoes when you put them in the ground. You will see that the main stem above the soil line on the young plants is "hairy" in nature. Roots will grow from that stem if planted deeper. Pluck off a few bottom leaves and plant the tomato plant deeper than the existing soil surface in the pot. This will increase the depth of the roots searching for moisture and nutrients. For those adventurous or involved in a tomato growing contest, you could also add some mycorrhiza from the Greek *mykos* meaning fungus and *riz* meaning roots. This involves a symbiotic, loving relationship between a naturally occurring fungus and the roots, which extends the roots adventitiously into the soil profile to maximize their reach.

Remember Devo the American new wave '80s rock band that wore flower pots on their heads? At least I mistakenly thought they had a proclivity for plants and gardening. Instead they were iconic red-terraced "energy domes," designed to recycle some kind of wasted energy that flows from a person's head. In other words, the hats encouraged negentropy, which is reverse entropy or anti-entropy. It means that energy is recycled into good use and things become more orderly. Remember the "entropy" in our lives that you read in chapter 7? The definition of entropy is in part "a lack of order or predictability; gradual decline into disorder or chaos." I want some of the "flower pots" that would bring some negentropy to my planters. One way to do that is to start at the center and work your way out ... focal, filler, edger, trailer. Thriller, filler, spiller and you my friend are a rockstar.

We shouldn't fritter away our dopamine. I have experienced that dopamine is responsible for a runner's high. Physical exercise is one of the best things you can do for your brain. Whether gardening, running. biking, hiking or swimming, I believe it boosts development of new brain cells, slows down brain-cell aging, and improves the flow of nutrients to the brain. It can also increase levels of dopamine.

> Good intestinal flora is linked to a healthy mind, and who doesn't like flora right? You too can be an expert flora arranger.

So, I began exploring what vegetables and fruits would best encourage dopamine synthesis in my brain, combined with exercise. Don't waste your dopamine, enhance it without the use of drugs. I learned a healthy gut was the first step. Yogurt and fermented vegetables like sauerkraut became an important dietary habit, because there is a direct connection between the health of my digestive system and my cognitive abilities and mood. Good intestinal flora is linked to a healthy mind and who doesn't like flora right? You too can be an expert flora arranger.

After that I explored what vegetables or fruits would best improve my dopamine production? I am not a people doctor. I am a plant doctor. I developed my list of favorite nutrients, that I believe are somehow and some way linked to dopamine production in my brain. I have no scientific evidence, but if you want a personal kick in the plants, I recommend adding these to your diet:

     Avocados

     Broccoli

     Kale

     Brussel Sprouts

     Bananas (in my smoothies) I hate the texture.

     Almonds

     Beets

     Turmeric

     Ginger

     Tomatoes

     Any kind of Beans

     Coffee

     Oregano

     Cooked Oat Bran

     Cherries

     Salads, Salads, Salads.

And last but not least, even though we don't grow it in our gardens....Dark Chocolate.

That's my story and I'm sticking to it. Lettuce party like it's 1999!

Instead of a futile rain dance
With his sprinkler he'd take a chance
Drip turned to monsoon
Creating a lagoon
Don't look now but I just wet my plants

Chapter Ten

I just Wet my Plants

WHY DO WE SOMETIMES DO THINGS WHEN WE KNOW WE shouldn't? We know better, but it's almost like there's a mental override or block, as though our mind went on screensaver. Later we ask ourselves "What were we thinking?" It's ironic that distance at times seems to make things clearer. It's like the adage "you can't see the forest for the trees." Sometimes we're just too close to see clearly.

One February evening I was at the beach alone to take winter sunset pictures. When you become addicted to photography and sunsets, you are constantly in search of the big one, the picture of all pictures. It's like fishing, you don't want to bemoan the big one that got away.

Walking out on the ice in winter on Lake Michigan can be very dangerous. I do not recommend doing it and tell people not to do it. During the polar vortex winters of 2013 and 2014, almost the entire lake froze over and it froze like a brick. When not completely frozen, there is a temptation to approach the edge where water meets ice. The water freezes to an imperfect surface with hills and valleys and dangerous crevices. If you go in the water, hypothermia quickly sets in, and getting out without help is next to impossible. You endanger yourself and anyone who would come to your aid. Don't do it.

On this February night the sunset view on the lake was spectacular. The ice had formed along the lake shore extending about 100 yards out. I couldn't position the shot I wanted from the shoreline, because the ice forming along the water's edge elevates to partially obscure the view from ground level. My better conscience told me not to, but I slowly, but surely, ventured out on the ice.

After I took my shots and was proudly congratulating myself, I began what would be a treacherous trek back to shore. Picking my way across the unstable bumpy surface, I realized that before taking the pictures, I was oblivious to my foundation, drawn out by the lure of the setting sun. Now, faced with the task of heading east back toward the shore the tundra felt mushy, unstable and precarious. No longer blinded by the sun, I could see the position I had put myself in.

As I gingerly stepped from spot to spot … it happened in an instant. One moment on a solid surface … the next instant in the water. The surface I was walking on collapsed like a guilty criminal under intense scrutiny . My instant impulse was to raise the camera above my head so it wouldn't get wet. Fortunately, I had made it far enough to shore, so when plummeting through the ice my feet touched bottom. With no one around I was alone and chest deep in the drink. Instantly, a freezing numb feeling of pain like knives jabbed my skin, tempered only by the adrenalin pulsing through my body to escape this fix. Sliding my camera across the ice, so I could use both hands, I pressed against what solid surface I could find to try to lift myself from the water. Finally, finding a surface strong enough to hold, I lifted myself from the water, weighted by the water-logged winter clothes and boots I was wearing. Inside a marsh of ice chunks, sand and freezing water, the sun had now set, and temperatures were frigid.

I realized getting out of the water was half of my problem. I still had to walk home in this condition, and decided my best bet was to keep moving quickly. It is amazing how you lose feeling after the pain subsides, and recognize that frozen is not a good condition to be in. It almost felt like a dream, as I realized I was both walking and running, but could not feel it due to my numb condition. It was like I was sleepwalking through a dream and intent on reaching my destination. It felt like one of those dreams, where you are trying to run away from something, but can't get your legs to move. I was fortunate to escape my dumb polar expedition, and eventually made it home to slowly thaw out. I learned a lesson from my adventure and grew from the experience.

I realize a landscape lives in the polarity of frozen dormancy versus growth … unless you're a cactus or a rock. We don't think twice of the impressive reality of moving from deep sleep to the waking experience and back every day. It's the stuff dreams are made of. Embrace the natural experience I say, and if your neighbors look at you inquisitively, you know they are realizing they are missing out. Sometimes you just have to go outside and wet your plants; it will entertain your neighbors, and your begonias will appreciate it.

When we are truly awake, we have a sense of time without the use of a clock. We are in the moment, the experience, the place … love the plot you got. When we are asleep there is a sense of timelessness or dormancy. I have met people who sleepwalk through life and are physically awake but dormant. They have foliage, but they aren't blooming. Sometimes they need a kick in the plants experience to wake them up. The polarity of a quiet, dormant, snow-covered landscape wakes up to a spring explosion of color, sounds and visual stimulation. Living on the shore of Lake Michigan, I am amazed at the power of the gales of November and the roar, roiling and churning of a massive expanse of water. Conversely, walk the beach in January, when the water is suspended and frozen, and it's like walking on the surface of the moon. Quiet, still, dormant, suspended and waiting for something to happen. It always makes me think of the multitude of people I have met in my life, who fit into one of three categories: those who make things happen; those who watch things happen, and those who wonder what the heck just happened .

I have come to realize, that, some people, like plants, from time

135

to time are dormant, because it's more comfortable for them. If an activity like gardening requires an element of creative thinking, some would rather just throw in the trowel. Or, have someone else do the thinking for them which takes all the fun and experience out of it. I believe creative thinking is natural, innate and ingrained in all of us. If you make yourself believe that you don't have creative thought, then your default solution is satisfaction with being dormant. Watch kids without adult supervision. They figure it out eventually without intervention. They innately are very creative and messy. Nature can be messy and inspiring at the same time. Kids are explorers, and for many adults that exploration switch gets turned off at some point.

> A failure is not the end. A success is not the end. Our landscapes and our lives are a work in progress.

When it comes to adventure and creativity in the garden, on the job or in life, I find people make three mistakes that quickly move them into default mode.

1. The fear of failure. Creative action and activity requires failure. We learn from failure, and we improve from failure. Go outside and kill some plants. Don't use the excuse "I have a brown thumb." You need to get out and wet your plants.

2. Don't envision a masterpiece akin to the ceiling of the Sistine chapel. You don't have time for it and you are not Michelangelo. Creative successes are often incremental, and simplicity can be inspiring. Your creation doesn't have to change the world, it can be a variation on a theme.

3. A failure is not the end. A success is not the end. Our landscapes and our lives are a work in progress. Nothing stands still or is static. Even concrete walls like the Berlin wall eventually come down. You are either moving forward or moving backwards. You and your

environment may appear dormant, but unless you are planted six feet deep, you are not standing still. Even a compost pile must be turned to be effective and heat up. My desk will always be a mess, and finding the phone sometimes requires it to ring.

Safe is good for pansies and petunias. Don't get your pansies in a bunch. If you want to gain some ground, you sometimes have to plant something outside your hardiness zone.

You are far better off approaching a landscape in bite-size pieces. Landscapers will complete an entire project, but they have equipment, plans, and it's not their first rodeo. For the do-it-your-self homeowner, they feel pressure in spring, because the crowd mentality presses them to act on the landscape in spring. There is a sense of urgency and the tendency is to bite off more than you can chew. The Japanese have a word for it called karoshi 'death by overwork.' Summer, especially Fall and even winter are a great time for some landscape projects. Gardening is not just a spring thing. Nature is a 365-day-a-year experience and so should be your garden. Not to work in it 365 days, but to experience and interact with it. Some of the nicest yards and gardens are established, planted, and rooted in time investment during the fall season.

You need to look for opportunities and seize them. Even the simplest of creatures are opportunistic. Bugs don't respect our boundaries, and what makes us think they will? The sap beetle is going to dive bomb your potato salad on the patio table, and the stink bugs and box elder bugs are going to move indoors for fall. Spiders will move inside with your houseplants, because your walls, windows and carpet are an ideal habitat. Japanese beetles are going to eat your roses. They, like the stink bugs, traveled all the way from Asia to enjoy the American experience. If you had traveled halfway around the world to get someplace, you wouldn't spend the day in a hotel room. Asian ladybugs will have you climbing the walls, because the house is simply an extension of their habitat. If you had the choice of living in a log and leaf litter or a climate-controlled house with wi-fi, what would you choose? You would be amazed at the number of insects that live indoors with us on a daily basis, and those who freak out at the site of an ant or spider indoors make me laugh. Bugs are naturally opportunistic, and we can learn from them. Granted, there is a difference between being opportunistic and annoying, but bugs are naturally industrious. Generally, I'll pick up an opportunistic bug indoors, not squash them, and move him outside providing it another new experience.

Regardless of whether you are defaulting to dormant or less than opportunistic, I like to think we are living our lives in seven-year cycles. It kind of gives us life chances in bite-size pieces. Changes in thinking, physical changes, relationships, and approaches alter in seven-year increments, from what I've read. I look back and I can see some truth to that. Changes and successes tend to come in those natural, seven-year cycles of life as a means of defining a natural ebb and flow in our natural growth.

Consider the Himalayan lily, *Cardiocrinum giganteum*. Just the name alone conjures up imagination. For many, the Himalayas is a very magical place. The Himalayan lily, true to form, grows to a majestic size the largest of all the lilies. In the garden world, it has a Mount Everest mentality when it comes to ascension, topping out with a nine to ten-foot stalk. Most of the time it exists as an unassuming clump of leaves. But every five to seven years it sprouts a tall flower stalk and blooms in spectacular form. The plant is not monocarpic, meaning some plants die after a dramatic flowering. The plant *Sempervivums*, known as hens and chicks or the agave plant, are examples of a monocarpic plant. The Himalayan lily, instead, reinvents itself every five to seven years and blooms where it's planted. With plants, as in life, some people leave us or die after blooming. Some are given the opportunity to continually reinvent themselves. That is why sharing our stories is so important. You don't have to be a mountaineer to understand that life comes in a series of peaks and valleys. When we have moments to exist as the unassuming clump of foliage in our environment, I consider those "basecamp" moments. As a friend of mine has reminded me, the best stories are told in basecamp and not at the pinnacle. Those stories and experiences teach me who I am and lead us to look more fully into what makes us who we are. Like the Himalayan Lily it's all part of the cycle of life, so we are well grounded and rooted when blooming time comes.

Don't let the fear of failure or the need for perfection keep you from trying. People want a definitive answer and schedule on watering and it just doesn't work that way. Plants are different, soil types are different, seasons change and exposures vary. You work towards answers. I was listening to an author who said, "you write towards the answers". You don't have all the answers so work towards the answers. Questions create more questions and that's okay. I can give a definitive answer to the watering question but it will be wrong at times. Thomas Jefferson said, "He who knows nothing is closer to the truth than he whose mind is full of falsehoods and errors."

Dandelions are considered a healthy salad delicacy by some, public enemy number one for others who are fastidious about their prestigious pitch. We get the name for this weed from the French in two forms. The more common dandelion '*dent-de-lion*' because of the tooth-shaped lion leaves on the plant. The second, in some parts of the world, an even more colorful word association of old French is used, *pissenlit* or pee in the bed. In some French-speaking parts of the world, including regions of Canada, *le pissenlit* is derived from the dandelions diuretic properties when leaves and roots are eaten. Some kids are told not to pick them as a Mother's Day bouquet, because they would make you wet the bed. During student protests and strikes in France in the 1960s, Charles de Gaulle reacted by considering the uprisings childish, calling the university students nothing more than "pissenlits." I'm sure this didn't go over well with the students who probably responded, "Chuck, weed need to talk." You may be turned off by their taste, but a healthy salade de pissenlits with a tasty vinaigrette dressing is good for what ails you. Remember, "Take me to your weeder" may have been the first words of the extraterrestrial in your backyard last night."

The beauty of containers is you can move them into the right environment or microclimate for your plants, and have a measure of control. Certainly, the quality of the soil and drainage capability is going to make a world of difference when it comes to success.

I had been watching my neighbor's containers that she'd planted with spring flowers, herbs and vegetables in early May. I could tell she was excited about them and tended to them frequently. She was out there watering all the time. As I casted a glance over there, I noticed from a distance that the foliage was becoming anemic. I could see from my yard that her geraniums were turning pale green and yellow. I didn't say anything other than neighborly pleasantries across the fence.

"How are you Amber?"

"Great. It's a beautiful day isn't it?"

"Perfect day for gardening."

I kept my two-lips shut on offering advice unless it was sought. Finally, by late May she peered over the fence with a concerned look on her face.

"Rick, can I ask you something?"

"Sure, what's the matter?"

"My plants don't look so good. Do you know what's wrong with them? I've been watering them every day."

Without hesitation, like a doctor on call, I jumped the fence on her invitation to take a closer look, hoping it wasn't too late that what I had surmised from a distance was evident up close. The soil was a waterlogged quagmire and the plants were drowning. Her plants had been subjected to a form of Chinese water torture continuous drip and were completely stressed out.

I lifted the poly containers and poly window boxes from their positions, and found the drainage plugs at the bottom. I pulled the plugs and water came pouring out.

> *"You need drainage holes in the pots?" she*
> *said inquisitively.*

"Amber, you just wet your plants."

Like my friend Jen would say, "If only plants could talk."

When we are first starting out or venturing into a new project, it is like the stratification process for seeds. Stratification is a pretreatment, mimicking the conditions a seed must endure in nature before germination. They call it an "embryonic dormancy phase" and sprouting just isn't going to happen until dormancy has ended. Often seeds are subjected to some chilly temperatures or moisture that mimics nature's pattern. We, like seeds, all go through some form of stratification process, with some needing longer than others. Without it, we can't naturally bloom where we are planted, and nature always needs to take its course. You can't break dormancy from your current position if you don't understand your personal stratification.

Now, scarification takes it to a whole other level. With scarification seeds have their hard coat nicked, damaged, marked, softened, cut, to speed and enable the sprouting process. Sometimes it takes just a good soaking; other times they need to be roughed up a little bit. Isn't that a natural metaphor on life? How many times have you been nicked or cut to enhance your blooming process? Hopefully you haven't had to endure the "evacuation" method of this process. Some plants are "colorful" to attract the attention of herbivores or the avian species. Their consumption of said plant, results in seeds traveling the digestive tract and evacuating out the rear end, now marked and softened to sprout into something special. It's a dirty job but someone's got to "doo" it. If you've been crapped on you have to stand up to it and bloom where you're planted.

It might be embarrassing, but sometimes you just have to wet your plants. You can be frozen and dormant or abundantly adventurous. If anything, it will entertain your neighbors, give people something to talk about, and give you some great stories to share. The key is to find that sweet spot and just wet your plants.

The vine had become migratory
And invaded her territory
Her space she conceded
Intervention was needed
Can you tell me the story Morning Glory?

Chapter Eleven

Da Vine Intervention

I WAS SITTING IN THE KITCHEN, ENJOY-ing my cup of coffee on a cool but beautiful Dallas Saturday morning. Listening to the Neil Sperry gardening show on the radio, I was looking across the back patio at a ten-foot tall hedge my host for the weekend had asked me to trim. The hedge needed a serious setback. The disheveled hedge row was out of hand and needed someone to show it who was boss. I was assigned the job, but it would have to wait until my coffee was done.

The hedge was *Nandina domestica* also known as "sacred bamboo" or "heavenly bamboo." In my humble opinion there is nothing heavenly about this shrub. In winter the foliage is tinged red, in the south, where the plant is considered by many as invasive. Up north, the size is limited as a marginally hardy zone 6 plant and is deciduous. As far south as Texas, the hedge will retain its winter foliage, acting almost like a broadleaf evergreen with a red cast to its winter coat. For homeowners in Texas neighborhoods, it would be a rugged and aggressive grower, perfect to delineate lot lines and provide some backyard privacy. The root system is thick and almost impenetrable, and hacking at it is like maneuvering a jailbreak with a pick axe and small shovel. Well rooted, the upper portion quickly grows to the size of a department store, requiring said homeowner to live with their commitment while applying tough love.

I was assigned the task to "help" with this project but was in no hurry to acquiesce. Topping off the coffee, I decided to wait for the February sun to come up over the hedge, before heading outside in the chilly morning air.

Sitting at the table I heard a boom and the house seemed to shudder. I sprang to my feet with the sound of the clatter, and headed to the door to see what was the matter. I remember stepping out on the patio and looking at the roof from various angles, because it felt and sounded like someone was on the roof or had jumped on the roof of the house. I didn't see anyone on the roof or anything else for that matter, so I went back inside.

A short time later the radio show was interrupted by a special news bulletin. The boom I had heard, and the slight shudder I had felt in the windows, had been a sonic boom … a sonic boom created when the space shuttle, Columbia, exploded upon reentry into the earth's atmosphere. It was February 1, 2003 and I had been audible witness with a front row seat to the tragedy and loss of the brave souls on that spaceship high above Texas. As the reports and developments were reported throughout the day, I remembered the sad feeling I'd had just like the Challenger tragedy back in 1986. Having been an Apollo-age young boy, my love for rockets and astronauts was natural and life-long.

The nandina would have to wait. It's gnarly vine like out-of-control branches would have to wait for "da-vine" intervention on my part, which now seemed meaningless and trivial. The community and nation mourned this national tragedy and prayed for divine intervention for the families, associates and country that had just lost seven brave crew members on the Columbia.

Everything slowed down that day and focuses shifted. Driving down the expressway, there were signs alerting the general public to report any debris found to federal authorities.

The debris would be critical in piecing together what happened that day, whether small or large. Over the years, approximately 84,000 pieces were found, making up about 40 percent of the spacecraft. In order to dig deeper into the tragedy, the public's help was needed in identifying parts spread out over a huge area of eastern Texas and western Louisiana. The FBI was involved and treated debris fields as crime scenes. If you don't learn from a tragic event; it makes it an even worse tragedy.

As Oscar Wilde would say, "Behind every exquisite thing that existed there was something tragic." There can be beauty in the life lessons that follow for us who remain when tragedy strikes. I remember times of loss and how everything slows down. Over time, you are able to see the lessons and the good that can sometimes come from loss and tragic events in our lives.

Now, I apply the life lessons I've learned to the pruning that takes place in our lives when it comes to setbacks. Whether a rambunctious nandina or an aggressive vine, we sometimes have to slow them down to get them to bloom. It is a natural order.

> *"Adopt the pace of nature, her secret is patience."*
>
> *– Ralph Waldo Emerson*

When a vine like wisteria, trumpet vine or morning glory refuses to bloom, we have to dig deeper. Often, they need some stress to reorient their priorities. Often, we take the wrong approach because we garden people are just too nice.

The phone call was from a nice-sounding lady named Mary Ann, who was almost apologetic for not being able to get her vine to bloom. Despite her best efforts of pampering, fertilizing, watering and friendship, the vine stubbornly refused to do its part. More than happy to produce copious amounts of foliage, it failed to meet its potential with some token blooms. The belligerent vine had her climbing the walls. She sent me a picture of her vine moving inside the house! It grew through an opening in the exterior wall and found its way into the heating and cooling ventilation ducts. It then did as plants do, it grew to the light finding its way through a heat register on the main floor and moved inside. She now had a houseplant, but still no blooms. It actually looked quite nice, growing across the carpet, while at the same time gave a Hollywood classic science fiction invasion of the body snatchers feel to anyone seated in the room.

These troublemaker vines need someone to show them who's boss. We can appreciate their propensity to improve themselves, moving up the ladder of success, but we need a little fun too. All work and no play makes wisteria a dull boy, if you know what I mean. A work life balance for a plant with a frondescence addiction. In most cases these nice ladies who call take umbrage to the stubborn plant but were too nice and needed an intervention. It was up to me to bridge the gap with some not so nice verbiage for their herbage ... to give them permission to get tough.

> There is nothing like a little adversity to make us hungry again and retrain our efforts.

A vine wants to grow and grow, sometimes in lieu of producing blooms. In those cases, some root pruning or stress will show the vine who is boss and get it to bloom. We, as people, are the same way as we climb the ladder of success. Success is not always a good teacher. Complacency sets in, doing what you've always done, and getting the same results. Some pruning, some setbacks and adversity reorient our priorities and our focus. There is nothing like a little adversity to make us hungry again and retrain our efforts. We are like woody vines when we stretch ourselves. It's not whether you're going to have setbacks, it's how you get up and respond to them.

With wisteria, as an example, you may find you have to apply "tough love" and at some point slow their growth to encourage blooming via root pruning. The process of a sharp-bladed shovel, punctured intermittently into the soil through the roots, provides the "set back" or stress the plant needs to slow down and fight for its survival.

An annual vine like morning glory or moonflower can use a similar lesson. Instead of root pruning, we can hold back on the water to the point of some stress, and the plant begins to flower. Flowers produce seed and the plant is naturally fighting for its survival. Future generations hang in the balance. In blooming, seed production will take place, and the plant has naturally ensured future survival.

In most cases, when sufficient sunshine is available, the vine will begin to produce blooms. Some stress, along with a boost of phosphorus availability, will get the vine to produce beautiful blooms. Many times, if vines are growing near a lawn area, they receive too much nitrogen. Nitrogen makes plants green and leafy. This is great for lawns, but not the intended purpose for our flowering vine.

I like to think that frost or cold-sensitive vines like *Ipomea*, a huge genus of flowering plants that includes the vines morning glory, potato vine *Ipomea batatas* and moonflowers, are a metaphor on life and management. You see, I think the best managers and coaches, the best partners in endeavors, have the ability to balance tenacity and sensitivity, to be sensitive and understanding of others, while able to be tenacious when needed. No bull in a china shop needed here; you can be tenacious and sensitive at the same time. The vines in the Ipomea genus are tenacious growers that bloom when dealt stress. They are also highly sensitive, just mention the F word as in "frost," in close proximity, and they wilt like a daisy in the desert.

As always some stress will reveal character, and, in the case of plants, show how well grounded they are. Plants with fibrous and shallow roots tend to be less adaptable or easy to grow than those with tuberous, thick or rhizomatous roots. When people complain of their vines not blooming or their houseplant's demise, the root of the problem is exactly that, the roots.

Take, as an example, a couple of houseplants that are the closest living thing to plastic known to man. The *Zamioculcas zamifolia* (ZZ) and the *Sanseveria* plant. Often seen in office environments or airport terminals, these two seem to thrive on neglect. Both have thick, waxy leaves and low rates of transpiration. You must, however, look at the business end of these plants to see thick roots. In the case of a ZZ plant, tubers allowing it to go long lengths of time without water.

Adaptable to harsh environments, the ZZ plant is native to Africa and found from Kenya to Zimbabwe to South Africa. Considering its native environment, you can see the plant adapted to long periods of drought in between wet seasons. For those restless, easily bored or with a great need to nurture; this is not the plant for you. Unlike a vine, this plant is slow growing and content to simply exist in a well-lit corner. Pest resistant and a stable household friend, homeowners tend to kill it with kindness. Kindness as in water, too much of it. They feel they must do something to demonstrate they care, resulting in a permanent rainy season for the plant. I recommend talking to it, or, better yet, get it a greeting card and rotate it a quarter turn from time to time. You'll both feel much better. It's an expensive plant to kill, because it is a slow grower. Because it takes time to get to size growers charge more for the plant. Similar to *Aspidistra elatior*, also known as cast iron plant, or *Rhapis excelsa* known as lady palm, their methodical slow growth make them durable but expensive for the homeowner. If you're just going to kill them with water, they might not be the plant for you.

Sanseveria is a colorful character known for its diversity, usefulness and lasting quality. The genus *Sanseveria* consists of numerous species and varieties with flamboyant, often variegated sword-like foliage, as flashy as the man they were named after. Raimondo di Sangro, Prince of Sansevero, was the world's most interesting man long before the advent of beer marketing, or at least Raimondo would tell you that. It is said he dabbled with numerous inventions, from super lightweight cannons to fireworks, hydraulics and waterproof capes and other bizarre experiments. He was the Prince of Sansevero and lover of knowledge, willing to testify of his prowess and fame. He would have been pleased to know that a Swedish botanist named the *Sanseveria* plant in his honor. A tough hard-to-kill houseplant, it lives as long as the legacy of Raimondo, provided you don't kill it with "kindness." A renowned office plant, ideal for indoor air purification qualities; its tough, dogged obstinance and resolve is reflected in its common names: snake plant, snake's tongue, devil's tongue or mother-in-law's tongue.

From ZZ plants to *Sansveria* and flowering vines, sometimes we must show them who's boss to get them to behave. Kindness is a virtue but sometimes too much just makes us soft.

Mary Ann could now envision blooms and fruiting that would rival the vineyards of Mesopotamia in years gone by. When in Rome do as the Romans do. The Romans understood the effectiveness of the pruning knife when it came to vineyards. My intervention helped her see the light, that her pampering and pleasant nurturing disposition enabled the vine to ramble without purpose, because she was too nice. She now could enjoy the days of vine and roses. Some, after intervention, claim they have added the method of a stern talking to in their repertoire of discipline. One caller said that a spanking worked, and she spanks the wooden trunk of her wisteria with a paddle spring and fall. I'm not sure if that works, because the wisteria may have enjoyed it. Regardless, a good chewing out to clear the air, along with a disciplinary paddle, will certainly make the human in this interaction feel better.

Remember vines grow. It is their "nature." You need to provide sturdy support structures with a tendency to "overdo it." I tend to overdo it when building something ... too many nails, too many screws, too many boards. I build for worst-case scenario.

Vines and vineyards have a rich storied history. From the Middle East to France and beyond, there is evidence of wine production, dating back to 4,000 BC and beyond, with numerous Biblical references celebrating vines. Talk about "Da-vine" intervention! I use grapevines on fencing in my yard, not for the production of grapes, but rather the aesthetic feel it provides in my landscape. The same can be said for Hops *Humulus lupulus*. With the rising interest in craft beers, growing hops not just for production, but their ornamental qualities, has become popular. Hops is an herbaceous perennial that is easy to grow. Provide sunlight and plenty of support, and by August and September the presentation of foliage and "nuggets" (hops) can be quite spectacular! Hops like a rich, well-drained soil, so till deeply with good organic matter for best results.

The list of vines you could try in your yard is extensive and fun. If you don't have a lot of yard space, going vertical may be just the answer to create an intriguing enjoyable landscape. Here are some vines I suggest you try growing in your landscape.

- Wisteria. Make sure to have a strong structure for this vine with plenty of room to grow. A wisteria can swallow a structure or building, but, when planted in the right place, the blooms and results are stunning.

- Clematis. The key to clematis is "cool roots and hot tops." Mulched at the base and with organic matter in the soil in a sunny area the flowers are simply gorgeous.

- Climbing roses. Heavy feeders that need sunlight and support. If given these three elements, a climbing rose can provide floriferous results for years to come.

- Trumpet vine. This aggressive grower is easy to grow, with some even labeling it "invasive." That said, if you have the room to grow, this woody vine produces blooms to attract Hummingbirds to your yard and provides a visual explosion. (Note that both trumpet vine and wisteria both may benefit from root pruning. See earlier mention.)

- Honeysuckle. With sweet yellow to orange and red blossoms, this easy-growing vine will attract pollinators, butterflies and hummingbirds to your landscape.

- Climbing hydrangea. With white lacecap blooms in summer and aerial rootlets, this vine is a "clingy" must in the aerial landscape with glossy green foliage.

- Hops *Humulus lupulus*. An herbaceous perennial that is ornamental in summer and has broad interest due to the craft beer industry.

- Passion vine *Passiflora caerulea.* For exotic blooms this might be your vine. Not hardy in Michigan (herbaceous habit surviving in zones 7 or warmer) but don't let it keep you from trying this vine for its intriguing blooms.

- Morning glory *Ipomoea purpurea*. What's the story morning glory? Easy to grow from seed in warm weather, this vine has adorned many mailbox posts and lampposts in its day. A tender annual, its tendrils are fast growing and adept at pirouetting. Moonflowers are similar in their growth needs. This tender annual vine has unique unfurling white blooms, perfect to be used on decks where evening entertaining will take place. With moonflowers, as well as morning glories, go easy on the nitrogen.

- Sweet potato vine *Ipomoea batatas* are wonderful annual vines for containers.

- For a tropical feel in your summer landscape, plant a *Mandevilla*, perfect for containers, fences or patios in a warm, sunny spot. I learned to play the mandevilla in high school (just joking.)

Knee deep in a muddy morass
It's hard to show some class
But determined to evolve
And with great resolve
No obstacle would form an impasse

Chapter Twelve

You are in for a Root awakening

WE ALL KNOW PEOPLE WHO ARE A STICK IN the mud. They're the party poopers, the manure in the punch bowl, and uninterested in change for the sake of being uninterested in changing. A **stick** in the mud trips all of us up with them. Someone **stuck** in the mud, however, might just need a helping hand, a nudge to move them from their current predicament.

One early spring day my patience with winter had worn thin. I was anxious to get started on a landscape project and decided to commence firing. The snow had melted and the drab canvas stretched out before me. The soggy conditions were not going to dissuade me from making progress. Besides, I had rented a loader and had the day off from work. I was no stick in the mud and I was going to show progress and get a jump on the season.

The property had a blue clay soil type that had veins of blue and purple colors. I quickly learned that this soil had a very high absorptive property. The soil can absorb water better than a baby's diaper, resulting in a huge increase in the soil's volume. Saturated from the melted snow the awaiting morass was ready to curb my enthusiasm.

With boulders at the ready and a stack of landscape timbers, my son and I were dressed for the occasion and ready to do battle. The goal was to raise the planting beds and define them with the border materials. We went to work, and, as we disturbed the earth slowly, the footing became a quagmire. The earth roiled beneath us, churning into a consistency of cake batter and soup. Our muddle quickly became an entanglement, as though we had awakened the earth from a sound slumber.

We had angered the loam gods and were now going to pay the price. Terra-firma had become a muddy mayhem and we were knee-deep in the morass. The loader quickly sunk to the point the wheels were no longer visible, and the mud encroached the seating area. With no choice but to eject from the pilot's seat, we now stood gazing on the equipment with a new priority. How were we going to dig out this steaming mass of metal and make it functional again? We formulated a plan to use the landscape timbers, under the wheels, to provide footing to back the vehicle from the muck and mire.

I have never seen such a sight in my life. The angered earth methodically swallowed the eight-foot long timbers, and they slowly slipped below the surface like a sinking ship. It was a "davine" comedy, as Dante's abandon hope all ye who enter here came to mind. I stood in awe as the mud-covered steaming hulk of invincible horsepower listed in the soup like a wounded vessel. We turned off the engine and stood in silent witness to what our hands had wrought. With just the sound of our labored breathing and the mist of our breath in the air, our surreal position was frozen in time.

I had experienced that helpless stuck-in-the-mud feeling once before, and my son was a participant that time as well. In that circumstance I was an unwilling participant. Soon after getting their licenses my son Rick and his friend Stuart decided to take Stu's car for a ride one March evening after church. Two 16-year-old young men decided that a shortcut off the beaten path was a good idea on a chilly moon-lit evening. The shortcut involved a corn field, which in March would be a gummy gooey sticky quagmire

154

of clay.

I'm sure once they got part way in they were committed, and, realizing the error of their ways, decided to gun it. This further enveloped them in their quandary until, buried deep in a corn field, the car was sunk to its windows in mud and in peril. After climbing out the windows, I imagine they stood in that cold moonlit cornfield debating their next move. Who are you going to call? At that point they went to plan B. If at first you don't succeed ask Dad to help you. I'm always plan B and the call was made for Dad to rescue the charlatan chauffeurs.

I arrived on the site with flashlight in hand trying not to laugh. I had to act serious with fatherly concern. Brown sludge dripped from the door handles, and their bespattered clothes and faces were evident in the light when I shined my light on them.

After my initial shock I composed myself and recognized a wrecker would be the only way to salvage Stu's car. Rick suggested a friend by the name of Justin who had a four-wheel-drive pickup truck. Against my better judgement I let them make the call, and soon the roar of the approaching rusty truck could be heard in the distance. The truck thundered onto the tundra and idled in the muck a few feet from Stu's car.

By this time the commotion and clatter had awakened observers in a nearby apartment complex who threw open the sash to see what was the matter. The area was farm fields that were quickly being developed into housing, and the early settlers now had a front-row seat to our dilemma. I saw them peering from the windows as though we were aliens. We were in deep so expedience would be prudent. After hitching the tow strap and chains between the vehicles, I instructed Justin to get in the truck and give it a go. The venerable, aging behemoth of rusty metal roared to life and shuddered as he stepped on the gas, spewing silt and muck for hundreds of feet, coating all those unfortunate to be within range. The end result was now a second vehicle buried, bogged down and steaming in the mire.

At this point Stu's Dad arrived on the scene. Two fatherly

heads are better than one. A crowd was gathering at the windows of the apartment complex as we were illuminated by the perfectly clear moonlit night. I imagine they had popped some popcorn and were watching for what's next with a scene reminiscent of a combination of *Woodstock* and the movie *Deliverance*. We did what we should have done from the start and called a tow truck.

As we waited for the wrecker my heart sunk. Off in the distance I saw the multi colored flashing lights approaching. These were not Christmas lights, the cavalry had been called by someone in the apartment complex and a police cruiser would soon join us. The ruckus had awakened all and the officer was about to visit our garden party. My mind raced as I was designated the spokesman for the group, and I wanted to avoid saying something that would aggravate the situation like, "Good evening officer, what seems to be the problem?"

The officer approached in his cruiser parking about 25 yards downstream and turned on his bright lights blinding us in the process. What a sight we must have been. I envisioned him stepping out of the cruiser and asking who's in charge here? Obviously no one I would say. Instead he stayed in the cruiser, and I wondered if he was calling in support or laughing hard trying to regain his composure. We stood shivering in the cold night air covered in mud, staring into his headlights in embarrassed silence, partners in grime.

For what seemed like an eternity, but was probably only a few minutes, he emerged from the cruiser as we saw a tow truck arriving on the scene. He didn't ask what happened or who was in charge here. Instead he did what police officers do, shined his flashlight in our mud-spattered faces and asked everyone to begin producing ID. It was at that point the cornfield adventure went off the rails. Justin was walking back to his truck to find his wallet, and, in the stillness, I heard a jingle of keys and a plop. I called Justin back and asked him if he had his keys. "No," he said, as we realized he had dropped them somewhere in the mire. The muck had sucked my dress shoes off my feet long ago so shoeless and cold we knelt in the muddle digging for keys while the police officer shined his light. I felt a tap on my shoulder and looked up. It was Rick asking if he could talk to me for a moment. We stepped back while the group continued the search for the keys.

"What's the matter?"

"Um, Dad we have a problem."

"What's that?"

"We don't have ID with us."

"What!"

"Our wallets are at home."

Envisioning all of us spending the evening in a jail cell to dry out, at that very moment lady luck finally smiled down on us. The tow truck was methodically pulling the mud-covered wreckages from the morass, Justin found his keys and a fortuitous moment unfolded. I realized that in the chaos the officer had locked his keys in his car which was running in the field. His attention diverted, I explained to him the vehicles were freed, and I was going to personally escort the boys home and send them to bed. Fatherly discipline was in order here, and we would leave the scene and no longer be a problem. He waved me off and we quickly climbed into our vehicles and drove off. Once on paved roads we dripped globs of mud for miles leaving a trail of memories behind.

Those memories raced back as Rick and I, now older and arguably wiser, stood next to the buried loader in my landscape. Knee deep in the mud, we lifted our legs to move as the earth sucked the boots off our feet and gloves off our hands into the bowels of the earth. The landscape timbers were gone forever entombed in their earthen grave. Years later, while we were planting a tree on that site, Rick was able to resurrect boots and the gloves from the scene of the "grime." They were geological artifacts of our historical failure that day, with the memories well-grounded in our minds.

FAILURES ARE TRULY GOOD TEACHERS. AT THE TIME YOU think you're being buried, when in reality you are being planted for future growth. I like to think the word future is a verb not a noun and that growth is an action. Sometimes messy, sometimes chaotic, judicious failures help ground you and move you towards better every time.

> At the time you think you're being buried, when in reality you are being planted for future growth.

Some plants instead of being planted are buried alive. Proper planting depth is a key to future growth. Roots need oxygen just like they need water, minerals and room to grow. I learned from those clay landscapes I worked in that you need to amend the surrounding soil with organic material for success. Proper planting depth and a plant's surroundings will encourage it to do more than survive, the environment will allow it to thrive. In clay soil people will often dig a hole and put some "good" soil in it. In essence, they have created a bathtub and the plant is doomed to struggle. Water collects in the hole, and the roots reach out horizontally and make an abrupt turn back towards the plant when they hit the walls of clay. If planted too deep, or, mounded with mulch, the foundation and environment choke out growth and vigor … an excellent metaphor on life. When I worked the entire surrounding area, improving the environment by working in organic matter with the parent soil in a 50/50 mix, the roots dared to venture into the soil profile. They could breathe and stretch into the soil, resulting in happier healthier plants.

Soil is something we often take for granted but it is vital to health. It seems not only plant health but people health too. I continue to read of scientists revealing that soil can have both antidepressant qualities as well as antibiotic characteristics. It seems dirt naturally understands antibiotic competition because it is teeming with bacteria. We continue to learn there is good bacteria and there is infectious or bad bacteria. In the quest for new therapeutic compounds, it's only natural one resource would be dirt.

Natural fungal partners in the dirt also play a role in the success of plants. We know that the texture of soil is important to plant success and incorporating lots of organic matter will usually make the homeowner successful. But in addition to organic matter it helps to have a "fun guy" around. Many people are not aware of a natural beneficial fungus called mycorrhizae, a thread-like fungal partner to plant roots. Mycorrhizae is in love with your plants, and helps roots extend into the soil, increasing root area, while providing a conduit for moisture and nutrients. Nothing new, mycorrhizae has been around through the ages doing its thing. It's like extending the rabbit ear antenna on our old TV sets from the '60s and wrapping them in aluminum foil for better reception. I remember those good old days. It was always frustrating, because when you touched the antennae the picture would clear, but when you released it the snow would come. Us kids would take turns holding the antennae. You can buy mycorrhizae to add to your soil as an amendment for better performance.

Root extension into the soil profile, to forage for moisture and nutrients, significantly enhances a plants adaptability to its environment. If you're well-grounded, both plants and people can weather changes to the environment or periods of duress. The goal is to thrive not just survive. It's no fun to be stuck in the mud.

This has been understood through the ages. The Greek philosopher Plato is credited with the quote, "People are like dirt. They can either nourish you and help you grow as a person or they can stunt your growth and make you wilt and die." That's really deep, Plato … real deep.

Stick-in-the-mud people slow everything down. Because they have made up their minds their rut becomes a risk-averse situation. Their roots circle their small space, choking out progress and inhibiting a flourishing existence. Spending time in their environment drags you in and slows your growth too. The key is to identify the current reality, then identify what is causing the stick-in-the-mud fixation and take it out, finally liberally adding in the amendments that allow all to thrive. The plot thickens.

When it comes to risk, environment and change, there is a lot of talk about climate change and global warming. Instead of debating the issue, I tend rather to adjust my approach to the gardening season here in Michigan. I plan to navigate these seasons each year in the following order from start to finish:

Dead of Winter

Spring tease

Winter returns

Dupery thaw aka Sham Spring

Third winter aka Spring surprise

Mud season and unfortunate frosts

Spring

Summer

Steam

Fake Fall

Summer part deux (usually a week or two in September when kids are back in school and you've packed away the shorts)

Real Fall

Holiday potpourri aka anything can happen

Change often comes from the bottom up. Sometimes we must look within as opposed to without. Sometimes we just have to look down first.

When problems develop in the landscape or with a plant, it is often best to look down, not up. This is contrary to our nature I suppose. We don't want someone to say, "Hi Bob, you look really down today." It is against our nature to look down.

You look things up. When I crossed the street, at a young age, I was told to look both ways. When you greet someone, you are taught to look them in the eyes. Motivational speakers tell you to look straight ahead, nothing but blue skies. When you're embarrassed you don't know where to look. You want people to think that you look like a million bucks. When I go to church I'm told to look up into the hills from whence my help comes, my help comes from the Lord who made the heavens and the earth. You look up definitions for answers. You look up websites, addresses, VIN numbers. You look up phone numbers, don't call us we'll call you. When you're worried you look out. Rarely are we told to look down.

Plants tend to mirror their foundation. If they are happy where they're planted they reflect vitality. Their foundation and their rooting is as integral to them as a foundation is to a house. Plants like an organic matter foundation with room to breathe and grow.

So, go ahead and give yourself permission for a "root" awakening. Help out someone who is stuck in the mud. You'll feel better for it. And remember the stick-in-the-mud people are sometimes mistaken for someone who is, as they say, solid as a rock because of their immovable position. In some situations this is necessary but not always a good thing. I've learned over the years….

> *Just because you're solid as a rock doesn't*
> *mean people don't wish you were a little*
> *"bolder."*

Remember Sisyphus was condemned to ceaselessly roll a rock to the top of a mountain. The stone would fall back of its own weight. Doing the same thing over and over again, expecting different results, can be a futile and depressing effort. Be a little bolder.

He never saw it coming
To others his life was humming
He needed direction
A course correction
The lessons learned were humbling

Chapter Thirteen

Staying Grounded

WITH HER NOSE PRESSED AGAINST THE screen door you knew what she wanted. Open the door a crack and she would slide through the opening and saunter into the garden. Coco the cat had a way of appreciating the nuances of a garden adventure. She would pretend to stalk her prey like a lion as she weaved between the ornamental grasses, hydrangeas and shrubbery. Coco could blend into the garden as her color was similar to a bale of Canadian peat moss. Searching for voles or chipmunks and then watching them scamper … was a game. Pretending to stalk she would never hurt anyone or anything. It was more like a cat safari and she was having fun. She would then sun herself on a patio stone or perch on the outdoor furniture, watching the world go by.

Coco was more interested in the *Calamagrostis acutiflora* ornamental grasses, more commonly known as feather reed grass, both the Karl Foerster and Overdam varieties. The cultivar name honors Karl Foerster, a German nurseryman, who discovered the plant in the Hamburg Botanical Garden in the 1930s. Coco's daily adventure to the landscape would involve time to snack on the *Calamagrostis acutiflora*. The end result was a pretty landscape with ornamental grasses that were completely flat on one side due to feline pruning. Strange-looking grasses, flat on one side, they became an annual occurrence in my garden.

Over the years Coco developed a long list of friends and admirers. When visitors arrived, instead of cat-like skepticism, followed by a dash under the bed to hide, Coco would make everyone feel welcome, and make instant friends interacting with all. Her internal motor would run, and her purring expressed the satisfaction she had in being around friends both new and old.

Some would say I am silly, caring for a cat the way I cared about Coco. I don't care. Coco wasn't just any cat. Some people don't like cats, considering their aloof and arrogant nature and skittish social traits as off putting. These characteristics did not apply to Coco who was friendly to everyone. It was always soothing to know I had a friend. After a long day at work she would bump her head against my leg, jump on my lap and purr and motor in contentment. Coco was expert at what they call head bunting, bonding her with others as a communication method of showing affection. When I would go in the basement and lay on the floor to do sit-ups or stretch before running, she would always follow and lay next to me pressing, her head against me as a form of encouragement.

Coco came along at a time when I was at the bottom and not well. I was struggling with depression and discouragement. Depression is a real condition that can strike anyone at any time. I always felt I was strong enough to avoid it, and optimistic enough to keep it from happening. A person in my position wasn't supposed to have these feelings and I just needed "to shake it off." I went through a period of depression, and didn't have a good handle on dealing with both the shame and the frustration. With a lot of responsibility, and, many people depending on me, I had to pull it together. I was being selfish, feeling the way I did, and needed to get my act together. I couldn't be me … I had to be what others needed me to be. It was at that time, exhausted and dealing with both physical and mental struggles, that I visited a doctor. He felt that getting a good night's rest would help make a difference, and he prescribed some sleeping pills for me. Zolpidem acted as a sedative and I would fall into a hypnotic, trance-like focus when under its spell. A good glass of Cabernet Sauvignon with a Zolpidem would make the worries of the day go away. I would take the drug and then fight it to stay awake. A conscious, hypnotic trance where no fear, worry or shame interfered with my activities. The drug is designed to deal with insomnia. I didn't care about that, it made the world co-operate for me. Frustration, disappointment and worry were no match for my little-white-pill solution. Why 'whine' about what is outside my control when I could 'wine' with my prescribed solution. It slowly spiraled into a problem where my personality suffered. A pill and a glass of wine became a night-time habit.

I had to make changes in my life. My personal issues were affecting those close to me, the business and my personal health. I had to learn to love myself and be myself, not what others wanted me to be. I decided that running would be my escape of choice. With the driven personality I have, I forced the issue and those early days of running were not pretty. It didn't help that I had decided to make this change in January, so night-time runs in the snow were more an act of survival than personal fitness. Initially, instead of a finish-line goal, it felt like I was running away from the demons that were chasing me. I guess that approach can help you stay the course when the training hurts. Never quit … you're not having a bad life … you're just having a bad day.

Slowly, but surely, the habit of running made me healthier and more clear-minded. I started entering running events, which were exciting opportunities to test my competitive and driven nature. More importantly with these events I learned something very important. In organized running events they don't move the finish line. Unlike life where the "finish line" is constantly moving and nothing is constant, I would be rewarded for my effort with a clearly designated and stationary finish. It is done. When running, as in life, looking back can make us tired, but looking forward builds anticipation. A lot of my depression was caused by looking back at my disappointments. Stress, guilt and shame are often self-imposed and become your enemy. Just because someone else is disappointed in you doesn't mean you should be. I think about that when out running, that God put two eyes on the front of our head and nothing on the back of our head. I need to move forward and not dwell on looking back at regrets. This is so important, that God gave us two eyes and only one nose and one mouth. This tells me that talking, eating and smelling are much easier than focusing. Never pass up the opportunity to be quiet. I get those opportunities when I run. I have an hour alone, where the only sound is the sound of my shoes striking the pavement, one after another ... moving forward.

Moving forward is the key. Moving forward is easier when surrounded by positive people. Negative influences in my life were dragging me down and I was allowing it to happen. I now understand how critical it is to surround yourself with positive influences. There is a direct correlation between a healthy, active body and a healthy mind. It "grounds" you. I love the quote by poet, writer and essayist Leigh Hunt, who grew up in London in the late 1700s and early 1800s, due to the fact his parents with loyalist sympathies had to escape the colonies during the War of Independence.

"The groundwork of all happiness is health."
– James Henry Leigh Hunt

I have watched other runners experiencing some of the mind games that I too experience in my head. It is a good lesson on life. The River Bank Run in Grand Rapids is a race in May that covers 15.5 miles over a 25K course. Runners are a fraternity of friends who meet at the start and finish line. They understand and accept the fact that people who don't run think we are crazy and a shoe lace short of a well-fitting shoe. In between the social aspects at the start and the finish you run a race with plenty of time to think. I don't bring my phone with me when I run and I don't wear earbuds, listening to music. I enjoy the silence of anticipation, as I listen to the rhythmic drum beat of hundreds of shoes striking the pavement, covering the first mile of the race.

For many the run to the finish line works something like this. At the start ... excitement and anticipation. What have I got myself into kind of thinking. A sense of uncertainty and twinge of anxiety for the unknown that lies ahead … some self-doubt that you prepared sufficiently … nagging concern that you should have visited the portable restroom one last time but the line was too long. The starting horn goes off, and a mass of humanity moves forward, funneled through a starting point in our individual journeys. The chatter of hundreds of people quickly drops off, and the silent rhythm of shoes on pavement set the pace. You measure your breathing and your pace to the others that surround you. The first mile or two is shoulder to shoulder progress, avoiding the legs of other runners while keeping pace. By miles three and four the pack thins out, and you settle into a comfortable rhythm and cadence, debating whether to pick up some refreshment at the initial aid stations or to continue on.

The key, by miles seven or eight, is not to allow self-doubt to creep in. The lesson I've learned is that in running, as well as in life's journey, when you have miles behind you and miles before you, the seeds of self-doubt creep in. When you're in deep, miles from shore, physical fatigue can start to creep in and play games with your mind. That is the time to train your focus. This is when the mind becomes more important than the legs. In the latter stages of the journey you question whether you can make it. You train your focus on those encouraging you in your journey, and, in turn, try to encourage others struggling along the way. Directing available resources to encourage both yourself and others is far better than self-imposed drama or self-imposed medication. Never quit. Press on. By the time you cross the finish line at mile 15.5, you begin thinking *that was fun* and *when do we get to do that again*.

Let nature take its course. Nature can be cruel and unfair but it is honest. The way of nature is honesty. Every time you are in the presence of unconditional love, you are in the presence of nature. While growing up, I was influenced by organizations where conditional acceptance and rules and regulations were the guiding principles. It caused a lot of stress for me and caused me to be very hard on myself. Those are lost years. Over time I learned it didn't matter so much what organization or list I adhered to, the guiding principle is very simple. Nature is simple and honest. Love the Lord your God with all your heart AND love your neighbor as yourself. The greatest of these is this. It's just that simple. We overcomplicate things, including nature. We mess it up. My church affiliation, professional affiliation, political party affiliation, race, sex, became far less important to me than my heart and soul. Group affiliations meant less to me, because group-think can be stodgy, inflexible, positional, and, most often … conditional.

Crowd mentality can often become "take a position and digs in your heels." The "membership has its privileges" thing. Defend and promote your position. Instead, I learned to love my neighbor as myself unconditionally. You, as a person, became important to me. I've done a lot of digging in my life. I've dug a lot of holes literally and figuratively. There is one thing, however, I dig less of today. The older I get, the less interest I have in "digging in my heels," and the less value I see in it. Digging in my heels can make me blind to my brothers and sisters. If an algorithm is a set of rules on how to solve problems, and, I only consider that set of rules, I am blinded to other opinions. That approach can dig in heels and cause one to surround them self with like-minded people who reinforce a position that might be wrong in the first place. Communication and media in today's world, is so fractured, with so many varied resources, that it is easy for someone to find a resource that aligns with their agenda. It perpetuates closed-minded thinking. I don't have to have the answers, and I would like to work with others to find the right answers. It felt to me that religion for the sake of religion was a list of what you should do, rather than what God has done. It was positional digging in my heels, instead of embracing mercy, grace and unconditional love. I moved my position.

It's hard to love yourself, because you know better than any-

> You have to be able to dig holes to stand on mountains.

one else the weeds that grow inside you. I read a quote that said, "If you became a person who could love unconditionally, everyone you love would flower before your eyes." That is as breathtaking as nature itself. Let heaven and nature sing. I don't think God or nature are thinking about what my church, business, or political affiliation is as much as God and nature care about my heart and soul.

WE ALL HAD 11 YEARS TO ENJOY COCO AND HER FRIENDLY personality. The day she passed away, I went for a long run, thinking about those 11 years and the person I was back then. I climbed up the back dune, with shovel in hand, and prepared a spot for Coco to rest. Under the shade of trees, with a great vista of the neighborhood she had enjoyed in her outdoor adventures … I dug the hole. Thinking of her friendly personality, I considered adjectives that would describe how I was feeling 11 years ago. I had felt some resentment, bitterness and even twinges of anger and vengeance at that time. Now, drawing close to the age of 60, I learn everyday how to deal with shame. Talking with other men, I find they, like me, tend to wear a mask that hides their shame. Now I try to reflect on regrets and disappointments in my life in a more honest way, acknowledging their existence and using them for good. It's not always easy. Love your neighbor as yourself. I feel no anger, no bitterness nor grudge against anyone. I'm not interested in head games. Don't carry a grudge. You don't have to be right. I don't have time for that … because life is short. I had a friend who once taught me that holding a grudge is like renting out space in your head for free.

As I placed her now lifeless body in her resting place, I thought about how Coco the cat was a friend to many, and how, in her simple way, she loved unconditionally in her journey. Imagine that. A life lesson learned from a cat who loved to nibble on my landscape.

His lawn was always meticulous
He groomed each and every funiculous
No weed dare exist
Within his midst
The time he spent was ridiculous

Chapter Fourteen

Lawn and Order

THE SUN EMERGED FROM THE clouds, warming the moist earth and awakening the senses. After thunderstorms the day before, and all the atmospheric nitrogen in an electrically-charged environment, the clouds now parted and the sun shined. The earthworms that failed to make it across the driveway, lay in their demise on the pavement, symbolic of the earth coming to life. The world was succulent and green … full of life. The robins probed the turf like a buffet, and leaves on the trees unfolded like the plot to a great story. The flowering crabapples and pear trees filled the air with perfume, intoxicating the busy bees. It was an early May day of promise and renewal … a great day to be alive.

I decided to leave work for my lunch hour and drive home. Wishing I could take the day off, I pulled into the driveway, peering over the dashboard at my front lawn. It was one of those days you could watch the grass grow, watch it springing to life, spreading faster than the latest neighborhood gossip. I could not resist the temptation, knowing I only had a short time before heading back to work. I decided mowing the lawn was more important than eating lunch. Instead of turkey on rye with a pickle, I opted to fire up the mower and cut the grass. With no time to change into other clothes, I proceeded suit, tie and dress shoes to begin cutting the front lawn. The backyard could wait until I got home, but the front yard was a reflection on my horticultural reputation.

Thinking no one will know, and that no one will see when you try to get away with something … everyone knows. At least it seems that way. People honked their horns and waved as they drove by. I remember thinking, shouldn't these people be at work? It's the middle of the day! Don't they have something better to do? Soon people were talking about how I cut the grass in my suit and tie, prompting the question what else do you do in a suit and tie? I guess it's a better reputation than being known to mow the lawn in my birthday suit. The al fresco homeowner, making a fashion statement while walking the lawnmower.

When you have kids, the lawnmower is a great appliance for finding lost toys and missing garden hand tools. Always in a hurry, I paid for it when failing to first inspect and walk the lawn before mowing. From branches to hand trowels to Hot Wheel cars, the mower hones in on foreign objects like a radar-guided missile. I'm lost in my own little world as the mower drones on.

Once, I was running my large walk-behind mower, deep in thought and oblivious to what was going on around me. The neighbor's boy, Reuben, snuck up behind me and pulled on my pant leg. I jumped what felt like a mile-high, startled back into reality, and I released the safety grip on the mower, instantly shutting off the loud engine. After expressing multiple expletive-deleted expressions, which I now regret, I clutched my chest as the engine backfired in a combustible belch and turned to the young man who said to me,

"Mr Vuyst are you cutting the grass?"

Impossible to get angry at this cute young boy but with my heart still pounding I said,

"Reuben, don't ever do that again. When you do that to Mr. Vuyst he says words that you will ask Mommy and Daddy to explain to you at the dinner table."

I've always had that problem mowing the grass. The warm air and drone of the mower puts me in a work-like trance, where my mind races in thought, and I'm in my own little world.

I have always felt that my mind races in a circular track. As I have learned from my studies on mindfulness the mind has a dual track. I am told that my mind tends to do only two things: it can recount and summarize the past, or it tends to think about the vexations of life planning, for the future. If my mind is always moving back and forth from the past to future planning, then I'm not in the present. Maybe that's why the drone of a lawnmower on a warm day and the mindless activity of walking back and forth is so good for us. The mindless activity makes us check out and be mindful in the moment, instead of racing back and forth from the past to the future and back again.

We learn the pattern of thinking in the past or racing ahead to the future as we move into adulthood because it helps us succeed. People who don't act that way are considered to be lazy or non-progressive. I think that's why we love flowers, plants or a walk in the woods. Plants are fully present in the moment. An almost Dominican principle approach when surrounded by plants or in the woods we "contemplate and hand to others the fruits of contemplation." The truth is if you can get your mind out of the dual-track mode for moments in time, you can better appreciate the nature that surrounds you. Time with nature moves us in that direction. The mind is usually bored in the present unless experiencing a stimulating occurrence. Nature can slow us down to be fully present, whether by a walk in the woods or laps behind a lawnmower.

Plants do not have brain cells and they don't think like us humans. An argument for wasting water on the lawn, when it doesn't need it, is that the roots will know to go deeper. In all the years I've chopped up grass clumps and roots I have never discovered a brain cell. They are, however, naturally "smarter" than you think, with a multitude of ingrained survivalist mechanisms. It is surmised that trees in the forest, with their interlocking roots, can communicate environmental changes or threats as a community. I believe it. A rudimentary but effective "web" of information. Remember, plants can make their own food, and I know a lot of people who can't cook or think they can cook but can't cook. Sunlight is a lot more than a cue or wake-up call for plants; sunlight is food.

You need a brain to process information. I read with fascination about plant neurobiologists who "believe that plants are conscious — not self-conscious, but conscious in the sense they know where they are in space ... and react appropriately to their position in space." They don't have brain cells or nerve cells like we do, but according to these scientists "they have a system for sending electrical signals and even produce neurotransmitters, like dopamine, serotonin and other chemicals the human brain uses to send signals." I truly believe plants can both send and receive rudimentary signals that allow them to react to their environment.

> I truly believe plants can both send and receive rudimentary signals that allow them to react to their environment.

All I know is everything goes nuts when the soil temperatures climb to 60 degrees plus in spring. I have had people write me, call me, ask me in response to GDD (growing degree days) comments I have made on the air; there is a fascination with the process. They are living it but not realizing it. Anyone who in spring has not carried a soil thermometer around with them has not truly lived life to the fullest. Probe the soil in anticipation of 60-degree ground temperatures and you, my friend ... are alive. We all live with the air-temperature forecasts projected by meteorologists to plan our week's activities with their seven-day forecasts. But if you want an earthy experience, however, learn to live by the ground temperature at the two-inch level, instead of the fleeting air temperature. And learn to live by the accumulated GDD growing degree days to anticipate blooming, insect activity and growth. You're truly living when you can take the low temperature for a day, add to it the high temperature for the day, and divide by 2 for a number greater than a base temperature of 50 degrees. If any number exceeds 50, you my friend have accumulated growing degree days, as in visible tangible growth. For example, on an early May day, if we have a low temperature of 39 degrees and a high temperature of 67 degrees, then 39 + 67 = 106, divided by 2 = 53, meaning we just accumulated 3 growing degree days. Glory, hallelujah! No wonder you feel so good. Unfortunately, 60-degree ground temperatures also wake up the sleeping crabgrass seeds overwintering in your lawn for germination.

Lawns are a way of taming nature in an age when urban dwellers are distant from nature and live in developments with association rules where they cut the trees down and name the streets after them. Lawns are a means whereby homeowners can map their territory. Now, understand I said map and not mark. Some animals mark via bio evacuation or scent or auditory squawk, which is unacceptable in the human, urban neighborhood kingdom. Lawns are instead territorial mapping of a sociographical plot, where habitual use becomes the home range, and horticultural prowess is put on display for all to see in the spirit of competition. Homeowners establish borders, as in their lot lines, and well-fed lawns delineate those borders, so confrontation is avoided, if you're really not into your neighbors. The lawn is an extension of themselves and will be defended if necessary. The call to arms is "get off my lawn" in the modern-day colonist vernacular of each neighborhood's one token grumpy curmudgeon. He's the king of his kingdom and needs to learn how to "get a lawn." Lawn live the king!

Don't be a crabgrass. That intrusive, meddling officious invasive vegetation that creeps its way into what was a good and healthy habitat. Bare earth is not a natural condition. Something is going to fill the space. The ground is bare the temperature rises and its invasive behavior is opportunistic. You can't be bare earth neutral. You have to stand for something rooted in your principles. It all happens so fast that quickly you are like the husk of a cicada with its legs in the air laying on its back on a sidewalk. What happened?

Crabgrass in essence is like negativity in the landscape of our lives. One negative person becomes two … becomes three, and soon you are surrounded by negative influences. *If negative people are not challenged, because they are difficult to deal with in the first place, their insidious behavior and opinion fill the bare space. Left unchallenged, their negative thought becomes, for some people, the truth even if they originally disagreed.* Slowly but surely the "crabgrass" fills the space and the environment changes. It becomes the norm. Positive people have a thick foundation of "turfness" to push back against the fear default of negativity or crabgrass and keep it from creeping in.

That's the life lesson here. Something is going to fill the bare space. Life goes on. Remember the feeling you get when someone close dies, someone leaves home or someone leaves your workplace, and you wonder how you're going to get along without them. Life goes on. You want the world to stop for a moment, to hit the pause button, but your reality is not their reality. Life goes on. Something fills the space. For those who remain, the challenge is to fill the space with something good. Don't let "crabgrass" fill that space.

Negativity is very different from behaviors that get results. From my experience in business, some people are charge-ahead, dominant types, wanting results and are impatient. Others use the pause button and are more patient in their approach, wanting more time or information. Is one completely right or the other wrong? Of course not, each has its place and contingent on the circumstances can perform at a high level. The same can be said for C3 plants, which play it cool like the Perennial Ryegrass in your lawn. They thrive in the cooler weather of spring and fall, and photosynthesize differently than the aggressive C4 plants like crabgrass that thrive in heat and use less water. Plants that use the C3 chemical pathway work best at moderate temperatures between 50 and 75 degrees. They lose efficiency as temperatures get hot. An example of a C4 would be corn, and who hasn't pulled up a lawn chair in July and watched corn grow in the Midwest? C3 plants can be become less efficient when the heat is on and water is less available. C4 plants thrive under the heat, and have a competitive edge in the hottest of days. Both categories of plants can thrive in the appropriate environment suited to their "personality".

The metabolic tasks plants employ is: taking in water and carbon dioxide, and using them with the energy of sunlight to make food in the form of starches and sugars. The plants thankfully exhale oxygen as a by-product of this process. People, like plants, adapt to their surrounding environment. People, however, do blame their metabolism for their physical issues, whereas a plant doesn't have that luxury. They simply adapt. That's why succulent plants are so popular. Highly adaptable, due in part to their fleshy leaves, they are the camels of the plant world that doting human owners tend to drown with kindness. Aside from their obvious thick sponge-like interior, that can store lots of water for extended periods, and their waxy coating epidermis or skin, they have a little secret few people know. Succulents and cactus use a metabolic pathway that employs *both* C3 and C4 characteristics, giving them amazingly efficient survival and staying power. They utilize a metabolism called CAM (Crassulacean acid metabolism) allowing them to take in carbon dioxide at night, known as evening respiration, and store it for use the next day. It is stored in cells in the form of malic acid, waiting for the light of the following day for use in photosynthesis. This is highly efficient and perfect for environments where nights are cool and days are hot. These plants adapt to their environment, and are a factory that works around the clock; they never take a vacation.

We can all thank Jan Ingenhousz, a little-known Dutch scientist from the Netherlands in the 1700s, for discovering the process of photosynthesis, a process by which plants convert light energy into fuel or food. I'm sure this thought or recognition of appreciation rarely crosses the mind of the homeowner as they walk the lawn cutting the grass. Jan, however, is credited with discovering that light was necessary for the process to occur. Light for plants is life like breathing is for us. I can go without food for days, but if I had to stop breathing? I can't go more than a minute without it.

> The difference between a mountain and a molehill depends on who is faced with moving them.

It's one of many reasons why trees and turf are not compatible. They don't like each other. Trees hog all the light from the turf below, and, worse yet, tree roots compete with turf for moisture and nutrients. Yet homeowners insist on trying to grow lush green turf under tree canopies. In the war between a shade tree and blades of grass who do you think is going to win? The turf, however, exacts some revenge when the homeowner damages the tree trunk with a weed whacker or lawn mower. All the "mower" reason that a thin layer of mulch or ground cover is a better option under the canopy of trees than turf.

The difference between a mountain and a molehill depends

on who is faced with moving them. A mountain can be moved, and, so can a molehill, but it is contingent on the level of determination you have. We already know that mountains are huge, and moles are determined and hungry, but tackling them is a whole other matter. Moles can dig at a rate of 18 feet per hour and can travel 80 feet per minute through existing tunnels, so, needless to say, they get a lot done in a day at your expense.

Many stories can be told of the adventures of mole control, and, from mishaps to missiles, they've all been tried. Many a frustrated homeowner has engaged in mole-vengeance tactics to protect their turf. I've interacted with people who sit in their basement filing the trigger on a choker loop trap to a precision edge. One show listener called in to declare he had named an elusive mole in his yard "Mole-No-More" Gaddafi after the infamous deposed leader of Libya. Another spent a fortune on smoke bombs, producing scenes reminiscent of the loveable but inept groundskeeper Carl Spackler. My Dad uses a homemade hose connected to the exhaust of his car, fished down the runway of the temperamental beast. Compressed air devices have been tried, and, when set off, create a rainstorm of bark mulch catapulted into the air. Sitting in a lawn chair, sipping on a beer with a pitchfork in hand, waiting for the earth to move is no way to go through life. The combination of pitchforks in one hand, and adult beverages in the other, as a hunting tactic has resulted in ER visits.

My friend Pete is one of the most creative, ingenious and mechanically skilled people I know. Pete is like the Professor on the *Gilligan's Island* show who can make a radio out of a coconut. Pete can mastermind a solution to anything that flies, rolls, moves, flows, explodes or exists. When Pete, however, told me of his contraption, using prods in the ground and electrical current to control moles, I warned him of the demise of a German pensioner outside of Berlin who killed himself trying to do just that. In the spring of 2009 we shared the story on air of a retired construction foreman who met his demise battling the lawn-wrecking rodents. Sadly, a police spokesman said the German gentleman was found lying dead in the garden of his weekend home, next to a 380-volt cable and metal spikes rammed in the ground. To add insult to injury, or, in this case death, the moles survived. The story that spring was the buzz of the gardening world, and created pause and reservation for all the Carl Spackler groundskeeper mole-hunting cowboys out there engaged in lawn warfare.

After 25 years of mole questions and answers, I don't want to engage in outlining mole control for your lawn in this book. I'm a grown adult and can do what I want to do. I'm tired of talking about moles. There are plenty of resources and options for you to tap into and choose the one that works for you. Just remember, explosives are generally not a good or practical option. In some cases, calling a realtor and moving could be a viable option. For my money, bait that mimics their favorite meal, earthworms, strategically placed in active runways, is a no-muss no-fuss effective means. May good fortune be at your footstep and be safe out there. Good luck.

Dogs can do as much if not more damage to a lawn than moles. Get a dog and there goes the lawn. It can be "ruff" to have a landscape that looks nice and is dog friendly. Can a dog and a landscape co-exist? According to the 2015-2016 APPA National Pet Owners Survey, 65 percent of U.S. households own a pet. It is estimated that 70-80 million dogs are owned in the United States, and that approximately 37 percent to 47 percent of households have a dog. With that many dogs I think we can learn to co-exist. Many are considering how to make their backyard environments pet-friendly and safe. This can have some "paw-sitive" effects for both dog and homeowner when it comes to lawn management.

Creative dog owners are setting aside a place for a dog to relieve himself using pea gravel or mulch chips, so they avoid the flower garden. There are romp-proof plants for your landscape like spirea or ornamental grasses that can take a licking and keep on ticking.

Running tracks, raised beds, and avoiding poisonous plants, as well as using pet safe weed controls and fertilizers, are of high priority to those with a canine companion. Ground covers, in lieu of lawns, can make the terrain both pet and human friendly too.

Some homeowners are giving their dogs a lookout platform to survey because dogs are territorial. Dogs like to patrol the perimeter of a fence line, so avoid a manicured turf along the fence; they like to parallel "bark." A designated, trained, digging pit might also keep him from ripping up your plants … if you frequent the pit with treats that he has to unearth from time to time.

I generally recommend picking an area for you, in a controlled environment, to have a patch of nice green manicured lawn, and also a patch for your dog to romp around in. For a new leash on life you can consider a water feature you can both enjoy on a hot summer day.

Can't we all just get along?

So, you're planning to do some planting? There is an old military adage that has been credited to numerous strategists that says, "No battle plan survives contact with the enemy." The enemy can be:

- when a plant is placed in the wrong spot and doesn't have enough space to grow.
- when there is not enough shade or sun
- when there is not enough air circulation and support
- when the soil won't encourage good root growth.

When any of these things occur, the plant can weaken, making it subject to attack from disease, exposure or insects. Take the time at least to determine what the projected eventual size of the plant is going to be. This applies to any plant, from the vegetable plants for your garden, the flowers for a container, the tree planted near a house or the houseplant meant for the living room corner. It may look innocent enough in that pot, but that plant may be a bundle of natural foliated branching opportunity and you don't want what I call post establishment "reset regret." At that point you might have to move it, or, worse yet … throw in the trowel. A plant, or a person for that matter in the wrong setting, is like brushing your teeth and then trying to drink orange juice. It doesn't work.

> A plant, or a person for that matter in the wrong setting, is like brushing your teeth and then trying to drink orange juice. It doesn't work.

Diversity is important, not only in our lives, but in the landscape as well. It is culturally important because monocultures become a negative when problems develop. The problem will spread like wildfire without diversity. This applies also when using native plants; we have seen native plants come under devastating attack and succumb when a monoculture is prevalent. Diversity is also aesthetically important for appealing appearance. Mix large leafed plants with smaller leaves, long grass like foliage with plants of horizontal habit, foundational evergreens with showy deciduous plants, woody plants with herbaceous, rough-textured foliage with smooth broad leaves. As the eye scans the landscape, it can rest on points of interest as opposed to a mish-mash of similar sizes and shapes. In a landscape of diversity, a continuous border of plants along the edge can sometimes mediate the variegation medley and get everyone to just get along.

Good plant combinations include these elements:

A. Different growth habits (upright, pendula, filler, spreading, etc.)

B. Different textures (glossy, ribbed, soft, leathery, evergreen, etc.)

C. Different leaf sizes: (narrow, wide, small, large, elongated, etc.)

D. Drifts or odd numbers

E. Good use of color

F. Differentiate between abundance and excess, allowing the eye a place to land.

SOME PEOPLE STRUGGLE WHEN IT COMES TO COLOR COMbinations and need a little help. I love to go to a paint store, and can be entertained by reading descriptive paint chips for hours, while, for others, the choices create paralysis. Allegedly, if you mix all the colors together, it will come out beige, and that's not very exciting. You need a plan Stan, and some basic principled thinking will color your world.

In nature, like a walk through the forest, large amounts of color in that setting tend to be similar, unless it is the fall season. Forests are shades of green with neutral browns and gray, and are types of environments that comprise analogous or similar colors, that for the most part are close to each other on the color wheel and lack contrast. They are beautiful in nature and relaxing or soothing. Colors opposite of each other, such as blue and yellow, will have vibration and play against each other in a dramatic way providing vibrancy. Two equal amounts of complementary colors create a positive, exciting and stimulating tension. You can soften the tension by having one less dominant in scope than the other to complement their colorful arrangement.

For those looking for easy, try a monochromatic color scheme. Just combine shades of a single color together. Choosing plants from a single-color family in different shades is pretty easy to do and not very dangerous.

Analogous colors are those that are next to each other on the color wheel. This is not very controversial as these colors tend to blend together well. Birds of a feather flock together.

Complimentary colors are opposite each other on the color wheel and will provide vibrance. Think of them as extroverts ... the life of the party.

Bright colors draw attention and make spaces seem smaller than they are, so reds, yellows and oranges would be appropriate, especially for distances. These guys are the extroverts at the party.

Dark colors or blue and purple create a calming atmosphere. In small gardens they help make things more spacious, and, combined with a running water feature, will make you subliminally feel cooler on a hot summer day. These characters are the introverts at the party, making an impact while flying under the radar. They are those with gravitas.

For those who can't make up their mind, or, don't want to be that strategic, a polychromatic mix of all colors is like the grand finale in a fireworks show without the noise. Those with organized mindsets may view it as a dumpster fire, but, if it makes you happy, you don't have to be sad. It will be unpredictably dramatic at the very least.

One of my favorite colors in the landscape is pink, because it plays well with almost any color ... even orange. Orange can tend to be overbearing, and too much of it makes a landscape look like a construction zone on a highway in summer. Pink, combined with orange, however, pops and is a marriage made in expression.

Finally, when it comes to color, don't forget the nonpartisan or neutral colors such as green, chartreuse, silver, white, brown and black. The personality of these colors is neutral, like vanilla, providing a happy medium. Can't we all just get along? Neutral colors are important, because they bring the personality out of other colors by being the canvas against which other colors are measured. Green can give confidence and support to other colors, and white helps brighten or lighten the mood of other colors. White is perfect for near outdoor entertaining areas as it will show up when the sun goes down. Brown helps make other colors appear well-grounded, and black makes other colors more mysterious and sultry.

There is a quote attributed to Dwight D Eisenhower, "Plans are nothing; planning is everything." Now I'm confused. I guess if we all just get "a-lawn" the world will be a better place, and we'll all enjoy a little lawn and order.

On his position he would not budge
In addition he held a grudge
through life he would trudge
As jury and judge
Content to slog through the sludge

Chapter Fifteen

How are your Gibberellins?

HOW ARE YOUR GIBBERELLINS? To a plant, gibberellins are important. Like humans, plants also go through hormonal changes. Gibberellins are plant hormones that regulate growth and influence various developmental processes in plants. Gibberellins are known as the speed hormone in plants, increasing stem length, improving fruit shape, delaying senescence (aging) and help a plant break dormancy. To me, gibberellins are the perfect metaphor to one's approach to life. In a plant, the gibberellins help the plant stretch itself, increasing the length between the internodes of a stem. I see those internodes on a stem like the years of our lives. The more we stretch ourselves the more we experience. We are more alive when we stretch ourselves, and the simile to plants is we delay our senescence. You are as old as you feel or think you are. Dormancy dulls our senses, but a good dose of gibberellins can wake us up like a plant in springtime. Now this may sound like gibberish to you, but dwarf plants can be sprayed with gibberellic acid to elongate stems. There is potential in all of us if we just stretch ourselves. I have a friend who always says, "I don't know what I don't know." I want more gibberellin plant-like moments where I find out what I don't know and stretch myself. Do you personally, like your dogwoods, look for opportunities to stretch yourself? Don't waste your gibberellins … branch out.

When it comes to growth hormones in plants I find the analogies to us as people to be endless. Gibberellin is a word I think of when I'm miles from a finish line during a race. I find myself stretching my resolve and endurance when running a long race. Another growth hormone perfect for analogies in my life is auxin. Auxin is a plant hormone produced in the stem tip that promotes cell elongation or growth. It has something called apical dominance, meaning it is found at the apex, the tip, the end of a stem. It works its way to the top. Here is where it gets really cool. Auxin accumulates on the shaded side of a stem away from the sun. Because it does that, it elongates the part of the stem away from the sun, making it longer than the part of the stem exposed to the sun. This causes the tip of the plant or the flower to bend towards the light! The cells on the backside of the stem elongate so the tip of the stem leans into the light. The obvious parallel here is, I imagine my auxin meter to cause me to be honest with myself when I look in the mirror and lean toward the light.

People react differently to stress and setbacks. Like humans, plants undergo a series of hormonal changes during a stressful period. I again often think of my imaginary gibberellins and auxin power when I experience a setback. In my mind, it makes me feel like I have imaginary superhero powers. Captain Auxin and his gibberellin powers! Production of gibberellic acid in a plant decreases during the stress of a drought. This gibberellin setback to a plant can affect its growth not only for the short term, but for the long term as well, resulting in problems years later. A stressed plant tends to mope, making it a target for disease and insects. That's why I try to surround myself with and associate with positive people. We are often a reflection of our environment. Whether we are plants or people, as a friend of mine reminds me....

Nature responds based on conditions.

The reaction in a plant to a setback when it comes to auxin and its apical dominance is a great metaphor for us as people. When a stem or branch is pruned or cut back, it forces outward growth.

Without the apical dominance, the stem now creates new branches, and full outward growth is encouraged. We all need a little pruning in our lives. Success is not a good teacher. Setbacks are great teachers and should force us to branch out. Pruning in our landscapes and in our lives is a natural and a good thing. Best-case scenario … we grow from it.

Prolonged periods of drought can be seen in plants in the short term as wilting, scorch or some defoliation. The longer-term damage, however, can be seen in dieback of branches, stunting of growth or decline in overall vigor and health. A plant's capacity to absorb water and minerals through the roots can be diminished as invisible root hairs are damaged. A reverse osmosis, so to speak, takes place drying living cells in the root, and their capacity is diminished. The analogy here to us as people, is how I have observed those who hold a grudge and the damage it does. Just like that tree that suffered a setback, it shows up years later in a negative way. I had a pastor and a friend who taught me that holding a grudge is like drinking rat poison and waiting for the rat to die. It's like renting out space in your head for free. It shocks me and catches me by surprise every time I see it in action. Out of the blue, years later, I have watched someone who, has at their own expense as well as the expense of others, withheld or undermined someone for the sake of an offense or event that occurred long ago. They carry it with them like a toxic photo album and can't turn the page. The fault is a seed that was planted and not weeded from their mind. Forgiveness is an important element of life. There have been plenty of times in my life that I've needed forgiveness. When I think back on those low points, the most useful analogy for me was a garden hose. Take a garden hose to the garden and set it on the ground.

Now turn on the faucet and watch what happens. Water is like grace. Water is like forgiveness. It always flows to the lowest point.

Gibberellic acid can make you smile by getting plants out of season into bloom. Most plants bloom seasonally, but a trained professional grower can get plants like *Spathiphyllum* to bloom by an application of the plant hormone. Also known as peace lily, some are more inclined to bloom than others. When I say, "how are your gibberellins?" it can also refer to your smile. You need a sense of "humus." It is organic to smile and laugh, and it's infectious for others. Maintain your "composture" and look at the sunny side of life. How many offices have I seen with a listless, languid and fatigued token peace lily parked in the corner collecting dust? The plant seems to reflect the disposition of the staff. Perk up your peace lily and pour another cup of coffee. We have work to do making someone's day.

> Water is like grace. Water is like forgiveness. It always flows to the lowest point.

Here's a little plant secret between you and me as it relates to your houseplants. Some houseplants are treated as prisoners of war from the time they are brought home. Subject to neglect, they merely survive instead of thrive. Some voluntarily take themselves out before their owners can kill them. With peace lilies people will say things like, "Get them root bound in the pot and they will flower for you." Or they say, "withhold water to the point of stress and they will bloom for you." There is some truth that stress will do that to a plant. I believe a tangible change in temperature between day time and night time temperatures is important, as that is what they experience in nature. And, of course, light levels are a factor and will often induce bloom. However, the plants and people analogy secret I have is this; plants will sometimes thrive and bloom just by a change in location. It doesn't seem to always matter if the spot is better than the last. The move just seems to get them excited. I have no scientific evidence, but it is something to try and won't cost you a dime. Sheryl Crow was right all along ... a change will do you good.

As I now move closer to my sixties, I think of how I had Lasik surgery on my eyes, my hair is turning grey, my body ages and my mind changes with experience, my diet changes, I take supplements, mistakes are made and fixed or forgiven, I dress differently, think differently, have different friends, tastes, opinions than when

I was younger. Am I the same man I was when in my twenties I began this voyage? Two schools of thought, no, that ship sailed a long time ago or yes, I am the same man just wiser and weathered. Am I fundamentally the same man or have I evolved into something very different?

I feel like the ship of Theseus. The Greek philosopher Plutarch makes a great point by asking whether a ship that had been restored by replacing every single plank remained the same ship.

> *"The ship wherein Theseus and the youth of*
> *Athens returned from Crete had thirty oars, and*
> *was preserved by the Athenians down even to*
> *the time of Demetrius Phalereus, for they took*
> *away the old planks as they decayed, putting in*
> *new and stronger timber in their places, in so*
> *much that this ship became a standing example*
> *among the philosophers, for the logical*
> *question of things that grow; one side holding*
> *that the ship remained the same, and the other*
> *contending that it was not the same."*
>
> *—Plutarch, Theseus*

In my mind I imagine Plutarch and noted Greek philosophers theorizing on the chicken and the egg conundrum, while lounging around the verdant scenery of the public gardens of Athens. If a grandfather's axe is passed on through the generations, after having both head and handle replaced, is it still Grandfather's axe? I have some old shovels from Grandpa in my garden shed where I have replaced the handle. If I give them to my kids someday is it still Grandpa's shovel?

When a chameleon changes colors is it still the same chameleon? All hydrangeas experience some color change as their flowers age, but some hydrangeas like big-leaf hydrangeas can change their color in a controlled way. It is easier to change a hydrangea from pink to blue, than from blue to pink. A pH adjustment can help do the trick, but not exclusively; it is more importantly the presence of aluminum in the soil. If you buy a pink hydrangea, and the pH of your planting area is acidic and there is aluminum present, which turns the hydrangea blue, is it the same hydrangea you bought at the store? Some people will throw aluminum foil, screws, nails, pennies or coffee grounds in the soil surrounding their plant. You're essentially creating a landfill and it's not going to change the plant's color.

> Life is going to be messy. Growth is usually messy. It's often not neat and tidy for both plants and people. And people, like plants, come in all different sizes, shapes, and personalities.

Change is inevitable in your life, for your hydrangeas, and, yes, even for the ship of Theseus. These are dig-deep issues fun for reflection as you prune your roses, feed your begonias and mow the lawn.

Life is going to be messy. Growth is usually messy. It's often not neat and tidy for both plants and people. And people, like plants, come in all different sizes, shapes, and personalities. Put them out there in an environment of relationships with free will, and it's going to get interesting. The famous French actor Jean Rochefort was once quoted as saying, "To you who enters my heart pay no attention to the mess."

I think about all the relationships I have built being part of the running community. There are a lot of people annoyed by runners, and can't relate to why we would create that environment for ourselves. I have runner friends who are so competitive they make little engine noises when they run by you. I have other friends who can't relate, and, like only things about running that aren't running, like eating pasta and carbohydrates, comfortable flashy footwear and being cheered. One friend points out to me that runners have "chicken legs." That thought sticks in my head now every time I run in a race and makes me smile as I run. I now wonder if people think I have chicken legs. We drink pickle juice or pepperoncini juice for long-distance runs. I crave a shot of pickle juice after a long run. You say tomato I say tomahhto and can we all just get along.

Whether walking or running, if you want a visual of free will and priorities, take a walk through a neighborhood and look at the garages. I had a neighbor who had a garage so clean, so organized, so immaculate, you could have exchanged it for an operating room. He mixed colorful speckles in the paint that he used on the floor. The garage had a stocked refrigerator and a work counter that was so sterile, you could perform surgery on it. After mowing the grass, he would meticulously wash, detail and wax his mower before it entered the garage. It felt like I had to wear scrubs and wash my hands in his sink before I could enter his garage. Every tool was in its place and labeled. I've had living rooms that don't come close to the creature comforts in this place. I would get excited when I would reach into the couch and find a cache of change between the cushions. In his garage everything had a place and a purpose. Nothing left to chance. I would just peer into the door because I didn't want to take my shoes off.

Conversely my garage was a wreck. I kept the doors closed, because that's what you do when your life is in disarray. I now don't feel so bad about it. A recent study I read said that with approximately one out of four Americans the garage is disorganized to the point you can't park the car in it. Go for a walk and spot the driveways with parked cars and closed doors. I know what's going on in there. Your house may be neat and livable, but everyone has a junk drawer and a junk closet somewhere. For some guys the garage is their junk drawer. I didn't wash and wax my mower when I was done cutting the grass. I was just happy to find it and it was a bonus if it would actually start. Once or twice a year cleaning out the garage was like a treasure hunt. You felt good about yourself and found lost treasures in the process. There were times I wanted to just put a garage sale sign in the yard and say come and get it. I have a five-figure investment left outside to the elements just so I could maintain my keep, donate, sell and toss piles in the garage.

Maintaining a neat garage was a challenge for me, and not a place I wanted to invest my time. I certainly didn't want to expend any of my "gibberellins" on it as there were always bigger and better things to do. Once, upon backing out of the driveway in the morning to go to work, I realized one of the kids bikes was lodged in the rear bumper left in the driveway the night before. I'm glad I didn't get far down the road before discovering the odd noise or I would have never had a chance at Dad of the year honors.

I remember a low point for me in my garage management responsibilities. There was a wet rain-snow mix coming down outside on a cold autumn night. I dashed outside, wearing only my shorts to make sure the windows were closed on the car. I re-entered the garage wet and cold, tripping over the kid's bikes, basketballs, mower, snow shovel and shoes. Three kids can create quite an adventurous stockpile in a garage in short order. Getting them to bed on time and keeping them alive was more of a priority than neatly storing their toys. After tripping for the third time and stubbing my toe, I straightened myself only to come face to face with a possum, sitting on the workbench with his head in a bag of bird feed. He lifted his head and turned his attention to me with the creepy grin only a possum can manage. We stood there staring at each other in awkward silence for a moment. The possum then continued to dine on the bag of bird food. I shivered and left the door to the garage open going inside not telling anyone. I figured when he was full he would leave, and no one would know the difference the next day. I've carried this guilt with me to this day, and now that it's out there I feel much better. He wasn't there the next morning, probably because he was both nocturnal and the bird feed was gone. He had to have had a stomach ache the next day.

I question where I would be today if I had invested time in my garage. Would I be better off or more successful? Would I today find satisfaction, if, I in fact had it all together? I've closed the door on that chapter of my life. I have lots of questions in my mind that I keep to myself. They float around in my head, and in quiet moments I have great conversations with myself. Meanwhile, people have asked me their questions for years. I've answered so many questions for others through the years that it seems more a methodical practice than an insightful exercise. Why won't my hydrangea bloom? Where is the bathroom? How do "we" fix that? Why is that making a noise? Why did that happen? What's going to happen tomorrow? Is the sun going to shine tomorrow? Why is my tree dying? When can I put down grass seed? There are days I feel my plot in life is the question-answer guy.

I reached the point that I cringe when someone opens their mouth and the first words are "I have a question "for" you." It's like they are presenting me with a gift. They have something "for" me and I'm about to unwrap their gift. I suppose we should be grateful for the questions in life and that others look to us for answers. I just don't want them to tell me they have something for me.

> It's not my job for me to be right, it is to **find** the right answer ... to **get** to the right answer.

Through all the years of questioning, I have learned this very important fact: I don't have to have the right answer or THE answer. It's nice when I have it, but it's not my job for me to be right, it is to **find** the right answer ... to **get** to the right answer. I'm not talking about an internet search for an answer. It's about walking a mile in another person's shoes. It's about sometimes dwelling in a mess and not having all the answers. It's about together finding RIGHT answers, not THE answer.

I have a note on the door of my office that says "the sign of true intelligence is being able to hold two opposing views in your mind and still be able to function." Over time, the questions change but so do the answers. I guess sometimes it's about having a messy garage and being okay with it. Maybe my next home should have a two "karma" garage.

The sage said learning is not by chance
Through study knowledge transplants
Perpetual learning is prudent
You're always a student
Now you're a smarty plants!

Chapter Sixteen

You can't go....
all the plants are going to die

(Don't throw in the trowel)

WHEN JOHN WINGER (BILL Murray) in the classic movie *Stripes* pleads with his girlfriend not to leave he says "You can't go, all the plants are going to die!" She slammed the door in his face, changing Winger's life as he enlists in the army and gets to know Sergeant Hulka, also known as "Big Toe." When my daughter Angie was producer of the show, she started the habit of playing that exclamation at the end of the show and it stuck. If you struggle with your philodendron or your pothos is pathetic, I wouldn't suggest something as drastic as enlisting in the army. It did reinforce, however, the weighty responsibility I had to babysit the plants of thousands of listeners each week.

One good reason for having kids early in life is, so you have someone to babysit your plants when you go on vacation later in life. They can assume the weighty responsibility of keeping your weigelas watered and your hydrangeas hydrated, while you gallivant off in some exotic destination. In addition, you will have someone to inherit your heirloom Christmas cactus that has been passed down throughout the generations. If you were stuck having to care for your grandma's mother's cactus, then you don't want to deprive your kids of the opportunity. This weighty responsibility usually comes with the inheritance of the family heirloom fruitcake to be re-wrapped for holidays to come.

I do love how the most recent generation of young gardeners are into naming their plants. Not botanical, scientific or common names. No these are relationship nurturing friend nicknames which is a dangerous practice. If you name the plant and it dies, it takes the burial in the compost pile experience to a whole other level. You've developed a nurturing relationship. You might have to call in sick to work that day. If Herbert dies or Succ-cute-lant succumbs, it's going to be a rough day. When Phil-O-dendron or Leaf Erickson again become organic matter, it can take a while to get over it. Think about it. We name our pets so what's wrong with naming your *dieffenbachia* Delores?

When I was younger people would find me odd for talking to my plants. I still stand by that principle, and not just for the beneficial carbon dioxide they derived. I think the vibrations and comforting tones of the voice of a "frond" indeed encouraged them to do better. I don't have scientific evidence to back this up, but I know at least I felt better. I also had free time in my garden because my wary neighbors kept their distance.

I read that the World Health Organization predicts anxiety will be a leading health issue in the coming years, outranking other issues like obesity. Much progress has been made in the past decade by many good people and organizations to make mental health no longer a stigma. Plant-based diets and an environment of plants can affect our mental health positively. It is well documented that the presence of plants can lift the mood of the office, the spirits of someone ill, and the disposition of someone sad.

I've learned in my life that some stress in our lives is a good and natural thing. It makes us alive as it causes us to negotiate our course in a fast-changing environment. We react to conditions in our environment. Plants react to stress too, but don't have the options we have at our disposal. Without a fight-or-flight option they can't punch or run from the source of their stress. Some plants grow so fast, they require a new zip code by the end of the summer, but they're still rooted and can't run.

Plants don't have a central nervous system like we do. They do, however, react to stressful situations like drought. They close tiny pores in their foliage called stomata, and, in some cases, fold their leaves along the midrib to conserve moisture. Plants can produce volatile chemical signals as warnings when attacked by insects. A rudimentary form of language appealing to rescue, such as the airborne chemical compound we smell from a freshly cut lawn. Many find the aroma appealing. At one time I had a cologne called "freshly mowed lawn" which I have since discarded. People looked at me strange as they caught a whiff and its allure was only enticing to robins and small pets. Small dogs would lift a hind leg next to me if I stood in one place too long. The reality is this: the aroma is a chemical compound alert that the turf is suffering and under stress. It's a cry for help. Now I feel guilty and I'm sure my vegetarian friends will be thinking about it the next time they slice up a cucumber. Rest assured that plants don't feel pain. The best way to describe it is they do suffer but they don't feel pain. For that reason, they are apathetic and simply don't care. It is what it is and nature will take its course. It's up to you to change the course of history.

I read with fascination a study out of the University of Missouri that shows plants can sense when they are being eaten and send out defense mechanisms to try to stop it from happening. Their study was based on a caterpillar munching on a plant, closely related to broccoli, and other plants in the brassica family. Essentially, the plant doesn't like being eaten and creates defensive chemicals that can repel the offending muncher. Only makes sense to me. If I was rooted and couldn't run, how else are you going to fight or flight?

Learning to identify plant stress is a key to having a green thumb. Because plants get stressed out from time to time, learning their signals and paying attention will elevate your horticultural status. Plants show stress through texture, wilting, and through communication via the foliage (discolored, undersized). Weather conditions, nutrient deficiency, pathogens and insect attacks are most often the culprits. Plants are aware of their environment and do react.

There is a difference between the ways of nature and an inclination to nurture. You learn as an adult that you can't fix everything. Sometimes you just have to live in, with, and surrounded by imperfection. It is a balance … control what you can, respect what you can't. Wabi Sabi is an ancient Japanese practice as old as the 15th century, and teaches us to appreciate imperfections.

It relates to a young man tending a garden under the master's tutelage. The young man manicures the garden to perfection, but, before presenting it to the master, he shakes some cherry trees so flower petals litter the ground. That stroke of genius is considered a celebration of imperfection. It's not sleek, perfectly mass-produced perfect surfaces. It's the weathered look. I have that Wabi Sabi look with the lines on my face from lots of sun and years of worry. You're thinking, *wonderful, I now have an excuse for my languishing lupines*. I don't have to airbrush or filter my selfies in social media. I'm wearing the chic Wabi Sabi look. Marks due to the passing of time, as all things around us, including ourselves, are in the process of returning to dust. When a youngster tells me I'm older than dirt, I'll take it as a compliment. I have an appreciation for rough textures and minimal processing, the weathered natural look instead of plastic. Appreciating instead of perfecting is the way of nature. We think in our arranged and planned, smartphone on the hip lives, it is the other way around. Nature is just like life … often you can't change the reality but you can change how you experience it. Gardens, plants and land-scapes are meant to be appreciated … not perfect. Wabi Sabi is almost like letting go of what you think your life should look like and celebrating it for what it is.

IT'S LIKE THE IMPERFECT YET FASCINATING RELATIONSHIPS of lichens. I am amazed at the duress lichens will cause some people when they call into the show. They view lichens as a sign of death or decay and strategize to bleach it from existence. True they are found on stationary imperfect objects like grave stones or the north side of slow-growing trees, but what keeps us from appreciating their natural beauty? Lichens are an intimate rela-tionship of fungal filaments and algae, creating a natural mosaic of art. The fungus protects the algae from the mean world out there and helps collect minerals and water. The algae, in turn, can photosynthesize making food like plants do, and fix nitrogen from the atmosphere. In this intimate relationship it shares some of that with the fungus. It's like a quid pro quo, you scratch my back I'll scratch yours intimate relationship of organ-isms. Who would want to break that up? She lichens me, she lichens me not. In the plant world we call the re-lationship a mutual symbiosis because they both benefit from the association. There are thousands of different spe-cies of lichens, and, in some cases, it gets kinky. The relationship becomes a threesome of a fungus, a green algae and a blue-green algae. Sounds like a joke, "a fungus and two algae walk into a bar."

> Nature is just like life … often you can't change the reality but you can change how you experience it.

Lichens are sometimes called reindeer moss or Icelandic moss, even though they are not mosses or plants. Because they tend to inhabit woody plants that are under stress or slow grow-ing or in decline, people tend to believe that lichens are killing their plant. They want to wipe them out with a scorched-earth approach. Lichens are simply using the surface to make love not war. Live and let live. As a matter of fact, the presence of lichens can indicate a healthy environment with little pollution. Wiping them out is as sad as coming between Rhett Butler and Scarlett O'Hara in *Gone with the Wind*. William Shakespeare gave Romeo and Juliet their moment. As with Napoleon and Josephine or Antony and Cleopatra, the love of a fungus and an algae can be love at first sight and happens fast. Why would you want to come between them? You might be altering the course of history if you do.

All of us stop living someday. So do trees. All the more reason to stop wishing away our days and to start enjoying what is going on right now. Everyone around you is going to die too. Enjoy those close to you now ... not later. When you get a foil-wrapped poinsettia at Christmas, enjoy it while you can. Don't stress over how you're going to keep it alive or get it to flower again. Poinsettias don't like cold drafts, and I'm not talking a beer here. Getting it to flower again requires having it outside in summer and then bringing it in and managing hours of total darkness and bright light to get the bracts to color again. Throw it in the trash or on the compost pile. You've got more important things to do. Happy Holidays.

STRAIGHT AHEAD

LIVING ON THE SHORELINE OF LAKE MICHIGAN, WE EXPE-rience straight-line wind storms from time to time called a Derecho. Derecho is a Spanish word meaning "direct" or "straight ahead." When these storms come through, they uproot what you thought was secure and what you took for granted. Winds up to 100 miles per hour that blow straight, instead of the rotating tornado winds, can flatten stretches of trees and structures. I experienced that on August 2, 2015 as I watched the storm approach from the north. Trees that topple are poorly rooted without a good foundation, or are top-heavy like Colorado spruce, acting like a sail in the wind. Some trees are simply in the wrong place, and a storm sooner or later was bound to do damage. In the aftermath of a storm, we discover that many toppled trees were in decline without us knowing or recognizing it. When they fall, their decline is evident from the hollow trunks and dying limbs we had taken for granted. It always causes me to question how grounded I am, and how well I will stand up to the Derecho, bound to come along in every life.

Sometimes it's okay to just operate by the seat of your plants. There is a difference between planning and over preparing. Get your "crastination" right whether pre, pro or somewhere in between. Crastination comes from the Latin crastinus "of tomorrow" so find your sweet spot and go for it. I like to tell myself that the future is a verb and not a noun. We thrive by contemplating our prospects. A plant in the wrong place has bleak prospects. Position yourself advantageously where you can grow.

Creeping phlox is the perfect example of a plant that makes the most of its 15 minutes of fame. The plant is inconspicuous and unpretentious for 51 weeks out of the year. Some would call its matted, weedy look, ugly. It's ready, however, for when the sun shines the first week in May, to take the spotlight and run with it for a few days, exploding in color with thousands of bright flowers.

Be able to recognize an opportunity when it arises and then be ready to make the most of an opportunity when opportunity knocks. I also firmly believe that God puts people in your life at just the right time so that you can impact their life or conversely they impact yours. The key is to recognize what is happening before the moment escapes you.

A friendship I developed at the radio station was with Rick Beckett, an on-air talent at WOOD, doing a popular morning talk show with Scott Winters. Rick was a controversial guy, and often in trouble for something he had said or done. In his mid-fifties, he was not healthy, dealing with diabetes and a life-long problem with alcoholism that he was very open about. A slip and fall on the ice had caused him to use a cane to get around. Rick would always summon me to chat when I was at the studio, and loved to talk about his huge collection of CDs and music. From time to time he would call me late at night out of the blue just to talk about random things. It seemed strange he would do that, but later I learned why. He was gifting components of his collection of music and would give me CDs and boxed collectors sets he owned, some unopened and still in the wrapper. One night in mid-February of 2009 Rick called me late at night to tell me he had a gift for me. He knew that I was a fan of Jerry Reed and asked me to make sure and stop by the studio the next day. Sure enough, as he had promised, the next day he was pleased as punch to hand me Jerry Reed's greatest hits on a CD in mint condition. He made such a big deal about it and seemed quite amused that I was now the proud owner of his music CD 'The Essential Jerry Reed.'

I remember the phone ringing, waking me my from my sleep in the middle of the night. My daughter Angie was calling, and, startled, I tried to focus on the words she was saying. Earlier that Thursday night February 26, 2009 paramedics had been called to Rick's home after he called his girlfriend, saying he was having a heart attack. Rick died that night at home, and, Angie, who was program director for WOOD radio at the time, was calling me to report the news. Rick was 54 years old and … just like that he was gone.

The next morning when I got into my car I reached over on the seat and picked up the CD Rick had given me the week before. I unwrapped it and pulled the disc from its case, popping it into the car's CD player. I drove quietly down the road, listening to the songs and replayed track number six *Smell the Flowers* a number of times. *Smell the Flowers* is a song sung by Jerry Reed and produced by Jerry Reed along with Chet Atkins, recorded on October 25, 1971 and released by RCA records. Chet Atkins, an amazing and legendary musical talent, collaborated over the course of his career with a couple of my favorite guitarists: Jerry Reed and Australian guitar legend Tommy Emmanuel. I have seen Tommy Emmanuel in concert from the front row a few times and marvel at his talent as a guitarist. Jerry Reed was quite a legendary guitar player too, however, most knew him as a lyricist, poet, humorist and movie star from the *Smokey and the Bandit* movies.

Looking back, I think of all the people in and out of my life in the past 58 years. I vividly recall details of conversations with those who have passed on from this life, and the impact they had on me. With each in hindsight, I now see they were giving me gifts, memories, advice just days and weeks before they died. It's almost like a garden, at full bloom, in its glory, during October before the first hard frost; they knew the end was near. They were helping me pause to smell the flowers. I've learned as we age our frame of reference grows, causing time seemingly to fly compared to when we were younger. As I rolled down the road, the snow stopped and the sun peeked through the clouds. I played the Jerry Reed song *Smell the Flowers* a number of times that day, and, after Rick's death, it became a part of me still today, as a reminder of how every mile is precious, every success fleeting, and every day … a gift.

When you go, the plants aren't all going to die. They can flourish with the seeds you've sown, the memories you've cultivated, the groundwork you've laid, and the legacy you've planted. Take your cue from nature. Flowers, snow, sunshine through the clouds, fall color, soil warmth, spring emergence … can all **quietly, quickly and unexpectedly** change the world around you. Your time has come. May the forest be with you.

His humor not always apropos
It slowly but surely on you would grow
The seeds would be planted
meanings that were slanted
to challenge the status quo

Chapter Seventeen

The Frozen Pundra

SOME PEOPLE DON'T LIKE PUNS, because it clouds their view of order and the way they think things are supposed to be. I believe that is why they become so annoyed with word play. I beg your garden. Puns are a form of questioning and interpretation. Multiple interpretations of thought, words and information promote curiosity, which, in turn, promote creativity where new ideas and approaches are born. It's how to build a better mousetrap, where ideas are exchanged, resulting in mouse-to-mouse resuscitation. Puns can be unsettling, because they challenge what we think we know, and are often dismissed as nonsense. Instead, I say it is opportunity for creative and abstract interpretations, linking one idea to the next if you have an open mind. The result is a new creation or at the very least an adaptation. It is not a fig leaf of my imagination; this is the essence of creativity, useful in our daily landscapes. Landscapes, like words, are not monotonous or static, and can be viewed from a variety of angles. Pundits are like flowers; they appear out of nowhere and change the perspective in a colorful way.

One of history's best pundits was Abraham Lincoln. He

found stories, tales, jokes and linguistic tricks, along with a sharp wit, served him well. If anything, Lincoln's sense of humor was amazing, considering his years of war and personal tragedy. A friend of his remarked that Lincoln's wit, humor and story-telling was his way to "whistle off sadness." If you think I am simply "pun"-tificating to justify my approach, read on, plants like people also will approach situations from a different angle instead of accepting the status quo.

> The status quo is not just routinely accepted, but, rather, a creative response is chosen entertaining a new angle on reality.

Biologists have discovered that plants can detect neighboring competing plants and match a response to their presence. They try stem-elongation to try to outgrow their neighbors, shade tolerance where they learn to live with less light flattening and enlarging their foliage or avoidance by growing away from their neighbors. In each case, the status quo is not just routinely accepted, but, rather, a creative response is chosen entertaining a new angle on reality. I believe people who can hold two or more opposing views in their mind at the same time, see both sides of a coin, and still be able to function, are just like those plants. It is a means of creative response to situations and others by considering alternative meanings and reactions.

> *My neighbor dug a hole in my backyard. I thought he meant well.*

Whether you can appreciate puns or not, two puns have stood out for me. Thank you very "mulch" stuck from the start, and are the first words out of most people's mouths when they meet me out and about. The pun that created the biggest stir was I just wet my "plants." You should too; it will entertain your neighbors, and your hydrangeas are going to love it. Since the early days of its debut, the phrase has "groan" on people and has been used on roadside signs. Those signs of course are perfect fodder for a social media posting and it spreads like wildfire. I would say that "Zip up your plants," "Don't get caught with your plants down," and "Don't operate by the seat of your plants" were winners, but "Don't wet your plants" has a leg up on them all.

I love it when listeners to the radio show get into the act. Through the years a number of fans of the show have approached me with their pun contributions. One day Cal approached me, and I could tell by look in his eye he had a good one.

> *A slice of apple pie on the island of Aruba will cost you $3.75*
>
> *The same size slice of pie on Barbados is $4.50*
>
> *When you go to Jamaica your apple pie will cost $5.25*
>
> *These are the PIE RATES of the Caribbean*

IT'S BEEN 25 YEARS THIS SPRING OF DOING MY GARDENING talk radio show, a quarter of a century of fun, interviews, sharing, learning, perspectives and puns. I'm awaiting another gardening season after a long, cold, snowy winter out here on the frozen pundra. Life is a wordscape, and with them we paint pictures and entertain with theatre of the mind. Don't get your pansies in a bunch if this small sampling does not get your stump of approval. I do not want to shrub you the wrong way. Lettuce at least agree to disagree for this too shall soon grass. Happy trails to you … until we weed again.

BONUS CHAPTER OF THE PHRASES AND PUNS I'VE USED through the years on the *Flowerland* show:

1. Thank you very mulch
2. Two Fun guys and a Cracked Pot

3. Be a smarty plants!
4. It's not a fig-leaf of your imagination
5. Maintain your composture
6. Trying to stay grounded
7. Just call me Planta Claus
8. You can't plant flowers if you haven't botany
9. Mum-believable
10. Entre-manure
11. Entre-manureal savvy
12. Entre-manures across the rooted plain
13. I just wet my plants
14. Use your tulips and speak to me
15. Mow and behold
16. It's not root-tine
17. Com-plant department
18. I'm a Crop-timist
19. Hot Plants never went out of style
20. We've got the plant-swers
21. Do you plant-asize about a beautiful garden?
22. The things I pot up with
23. Flowers are scent-uous
24. I am easy to get a lawn with
25. People phlox to me
26. Don't get caught with your plants down
27. Bon Foliage my friend
28. Termite to lose weight…starts eating lattice
29. Don't get your Pansies in a bunch
30. Floral you do this bud's for you
31. Lawn live the king
32. Floral the world to see
33. Maple we can help you
34. Don't operate by the seat of your plants
35. May the florist be with you
36. It's a kick in the plants
37. Fly by the seat of your plants

38. Split your plants
39. Bend over and split your plants
40. I love it when you talk dirt to me
41. I'm too sexy for my dirt
42. I'm just being knotty
43. I'm highly mow-tivated
44. Zip up your plants
45. Given our stump of approval
46. Can't believe I ate the vole thing
47. Vole-slaw
48. Frozen Pundra
49. Raw naturale
50. Pots and Puns
51. Basil Instinct
52. Be a pro-grass-tinator
53. I'm having dirty thoughts
54. Not mulch fun
55. Weapon of grass destruction
56. This too shall grass
57. I like to think I have a lot of grass
58. Part of the working grass
59. Grass warfare
60. Is your yard lawn gone?
61. The Grass of 2018
62. How long has this been growing on?
63. Oh bee-hive
64. Don't be sod
65. It ain't over until it's clover
66. Mole No More Gaddafi
67. Ar-Mole-Schwarzenegger
68. Rick or Treat
69. Planthora of fun
70. Pro-seed with caution
71. My mother told me there would be daisies like this
72. There's mower to come

73. Was hoping you would turnip
74. It's good today but what about tomato?
75. Don't asparagus any of the details
76. If you carrot all
77. Lettuce do this
78. Weed need to talk
79. Plant today eat tomato
80. Jack O Lantern broken? Get a pumpkin patch
81. Lower Da Bloom
82. Grow-ti-vation
83. Mow-ti-vation
84. Mulchos grassious
85. Frequent flower miles
86. Mulch to do
87. Plant ahead
88. Lettuce talk
89. Bonjour Manure
90. Fan-Grass-tic!
91. Aster not what your country can do for you….
92. Florist Gump
93. Air Florist One
94. Air Forest One
95. Dear damage
96. Out here on the frozen Pun-dra
97. Peas on Earth good till towards men
98. Been there pun that
99. Kiss me I'm Iris
100. I moss make dew
101. Lettuce turnip the beet
102. I'll be there in a snapdragon
103. I'm Gladiola you're OK
104. Before or Aster?
105. Leaf me alone
106. In short it's trunk-cated

107. Pun-tificating
108. I Pine for you
109. Spruce Springsteen
110. Shovel ready stimulus
111. Are you shear-ious?
112. Ground rules
113. You're grounded
114. Staying grounded
115. I'm sending you to your bloom
116. Don't be held hosta
117. Hosta la Vista baby
118. Wet your plants
119. Pull up your plants
120. Plant one on me!
121. What the world weeds now is lawn sweet lawn
122. Raking up is hard to do
123. Plant Man!
124. Involuntary plant slaughter
125. Consider it dung
126. Don't get uprooted
127. I used to be shy but now I'm an extra-dirt or Extra-dirted
128. You're in for a root awakening
129. Root for the loam team
130. Put your best root forward
131. I see London, I see France, I see problems with your plants
132. I'm your fearless weeder
133. Blessing in Da Skies
134. Da Vine intervention
135. Go out on a limb where the fruit is
136. Find this A-Pollen
137. I'm Up-Beet
138. You've bean such a good friend

139. Don't beet around the bush
140. Peas and quiet
141. Sprout it from the rooftops!
142. Lettuce romaine friends
143. Happy trellis to you….until we weed again
144. Sir Plants a Lot
145. Mulch to do about nothing
146. Thatch all folks
147. Tilling it like it is
148. Soil your plants
149. Traveling at the speed of blight
150. Kiss my Aster
151. What's growing on?
152. Scratch that niche
153. Class voted me most likely to re-seed
154. Bug-Wiser
155. Noble Peas Prize
156. Gardening is a work of heart
157. Weeding by example
158. Take me to your weeder
159. Not tonight, deer
160. Deerly departed
161. Holey Guaca-moley
162. Stalks are down
163. Diversify your hortfolio
164. Who's your lawnlord
165. I yam what I yam
166. Stop in the name of the lawn
167. Traveling at the speed of ground
168. The long arm of the lawn
169. Shrub you the wrong way
170. No Shrub-stitute
171. A knew experience
172. I've developed a flowering

173. Flower the leader
174. A sense of Humus
175. Mulch Sweat and Shears
176. We're Root'n for you
177. Stick with me
178. Dodge the crop-i-rotzzi
179. Ants in your plants
180. Groundcovers are fancy underplants
181. This Bud's for you
182. Grow with the flow
183. It's grow time
184. Tailor your plants
185. Need a good swift kick in the plants?
186. Bush it to the limit
187. May the forest be with you
188. Don't lose your composture
189. Make you green with Ivy
190. Pessimists need a good kick in the can'ts
191. Living on the hedge
192. Soiled reputation
193. Don't throw in the trowel
194. Who wears the plants in your family?
195. Lawn Irritation
196. Fescue 911
197. Gardener who needs a shave: Harry Potter
198. Favorite flower is Pillsbury All purpose
199. The prestigious lawn firm of Trillium, Sage, Gall and Fungus
200. Manure. It's so nutritious plants can't understand why cows get rid of it.
201. There are no guaran-trees
202. Lawn-gevity
203. I fought the lawn and the lawn won
204. Making a lawn story short

205. Even the trees root for us
206. Fun and Veg-u-cational
207. Hedge-ucational
208. You need a little lawn and order
209. Don't be a Mow-ron
210. You're among fronds
211. Avant-yard
212. Be a cracked pot
213. Rake rattle and roll
214. Sage advice
215. Stalk radio
216. Don't rest on your laurels
217. I'm green with ivy
218. Horticultural ADD: Always Digging Deeper
219. We've groan on you
220. We're just mossing around
221. Impatiens are a virtue
222. Be a Baby Bloomer
223. The Tree Musketeers
224. Hoe, Hoe, Hoe
225. Okey Dokey Artichokey
226. Soil the airwaves
227. Dirty job, somebody's got to do it
228. Hope your compost pile has a rotten day
229. Call Fescue 911
230. Phil dirt wanted
231. We're not bush league
232. When it comes to weeds, you've got pull
233. A new mow-lennium
234. Here we grow again
235. Can you dig it?
236. Gardeners know the best dirt
237. Weed em and reap
238. We'll grow on you

239. A growing concern
240. Trees a crowd
241. Berried treasure
242. Unidentified flowering objects
243. Don't get stewed about your tomatoes
244. Dirt cheap
245. Weeding things out
246. Field of dreams
247. Garden hoe!
248. We don't soft petal our love for flowers
249. Growing strong
250. The Lawn Ranger
251. Heard it through the grapevine
252. On the cutting hedge
253. Begonia is not a foreign country
254. Club moss is not a vacation resort
255. Make some prune juice
256. Add some humorous to your soil
257. Partners in grime
258. By trowel and error
259. Ace of spades
260. We've got our ear to the ground
261. Chip off the old patio stone
262. Down to earth
263. Turf's up
264. When the going gets turf, the turf get going
265. Turf city USA
266. Like an owners manual for your yard
267. Get to the root of the problem
268. Underground intelligence
269. Let it out or sweat it out
270. What's the story morning glory?
271. Hedging your beds
272. Barking up the right tree

273. Deep in their roots, all flowers see the light
274. Make a lawn story short
275. Can't we all just get a lawn?
276. Making molehills out of mountains
277. Well grounded
278. Well rooted advice
279. Don't be stumped
280. Giving you the silent tree-tment
281. Compost happens
282. Flowers going to pot?
283. Flowers a basket case?
284. Feeling bushed?
285. You need good dirt
286. What's the scoop?
287. Can't leaf it alone
288. Planting ideas in your head
289. Crazy like a foxglove
290. Flowers are mortal, weeds never die
291. Grass grows best where you wish it wouldn't
292. Flower Power
293. Surf and Turf
294. Yardeners are down to earth
295. Weeds are crack addicts
296. The way to a green thumb is dirty fingernails
297. Dig in
298. Shiver me timbers
299. If your plants need more support than a trellis, try group therapy
300. When the wood chips are down
301. Our plants are made with home grown ingredients. No sugar, no fat, cholesterol and proven to promote clean air and good health.
302. Branching out
303. Shear madness

304. Going out on a limb
305. Real gardeners never grow up, they just keep growing
306. Your branch office for garden advice
307. Weeding out answers for you
308. If you don't grow vegetables, it helps to praise and admire the ones in your neighbors yard
309. If you find yourself talking to your plants, you need to get out more often
310. Turning over a new leaf
311. Helping you find pay dirt
312. Breaking ground
313. Gardening is hotter than compost
314. When elephants fight, it's the grass that suffers
315. Annuals die, they don't come back so you'll come back
316. Don't hedge on asking for help
317. In – tree – guing
318. Take a dip in the Hot Shrub
319. Use your Two-lips and speak to me
320. You're the boss, Applesauce
321. You're just dandy, cotton candy
322. Ex-seed your expectations
323. You're the berries, Maraschino cherries
324. Got im-plants dirt cheap
325. Share the Soil-ties with you
326. Turf-ific
327. Wearing my sun grasses
328. Beet–tiful
329. Sod–isfying your curiosity
330. Need some re-leaf?
331. Kinda dug that one
332. Where have you bean all my life?
333. Hedge – endary

334. Seed of light
335. Grass –inating
336. A Mole in One
337. Mole Vaulter
338. Gauc – a – Mole – Eee
339. Stay tuned…there's mower to come
340. It grew my mind
341. Too-Mulch-uous
342. I'm at the Elm
343. I'm Flowered
344. Been there got the tree-shirt
345. Un-be-leaf-able
346. Those who throw dirt are sure to lose ground
347. Relax in my Rick-liner
348. We've got plant-swers
349. I've been raking my brain
350. It's a Tree-t
351. It's Oak-K
352. Lawn is flexing its moss-cles
353. Our show is com-plant–able with your radio
354. It's simply stem-ulating
355. Give the deer a taste of their own venison
356. Push comes to shovel
357. You've been working too yard
358. You work yard for the money
359. No condensation
360. Moss–age therapist
361. That's cucumber–some
362. Spring it on, Baby!
363. Really lichen this
364. Matter of thyme
365. Grow Get 'Em
366. In my ele-mint
367. A–pear–antly

368. A Pun-dently clear
369. I Di-grass
370. He has a Violet nature
371. Plant-alizing
372. Plants need to go potty
373. Don't be a spore loser
374. Thought that would Flora Ya!
375. Auger Management
376. Bud–wiser
377. Get a Manure–cure
378. Com Post Office
379. Pistel and Stamen is not an 80's band
380. Take the most direct root
381. Go ahead, rake my day
382. Forgive me I'm only humid
383. I knew you had it Zinnia
384. You Rose to the occasion
385. I beg your garden
386. Starving ar-tree-st
387. Vincent Van Grow
388. Prestigious Lawn Firm
389. Attorney: Trillium Sage Gall and Fungus
390. Secretary: Mary Gold
391. Keep us composted on your progress
392. You work yard for the money
393. Underground, undercover intelligence
394. Yard-core gardener
395. "Not a carrot in the world
396. Moss-ter gardener
397. Sedges have edges, rushes are round, grasses are hollow, what have you found?
398. Know maintenance gardening
399. Peas on earth good till for men
400. Auf-weed-ersehen

401. Haven't Botany in a while
402. Oh my raking back
403. Ariva-Dirt-chi
404. A grass act
405. Plant some pun-tunias

To all the assistants and interns on the show who each affectionately received a radio personality name from me:

Rhoda Dendron
Phil O'Dendron
Mary Gold
Pete Moss
Phil Dirt
Herb Garden
Holly Unlikely
Ivana Rosegarden
Wilma Plantsgrow
Branch Eyeapoka
Stu Manny Weeds
Althea Later
Yvette D. Plants
Patty O' Furniture
Alonzo Green
Moe D. Lawn
Moses Lawnagen
Dan D. Lyon
Rose Hips
Sandy Soil
Clay Pots
Rusty Pruner
Pat E. O'Stone
Candice Grow
Daisy Petals

Les Moss
Nomar Frost
Doug A. Hole

Thank you very mulch my friends. Happy trellis to you until we weed again.

– Rick Vuyst

Epilogue

(Epic-Log)

A Definition of "Entremanurial" Savvy

I CLIMBED THE LADDER CARRYING A heavy bundle of shingles to the roof of a barn my Dad and I were re-roofing. I planted one foot on the roof, and the other on the top rung of the ladder for momentum, and regrettably pushed off. I ended up on the roof and the old ladder ended up in pieces on the ground. My Dad and I looked at each other without a word, and understood further steps would have to be taken. We were now stuck on the roof. After a moment of silence my Dad said, "let me search this side of the roof, and you search the other for a creative way down."

As luck would have it, I found a way down, and was soon on the ground looking up at my Dad on the roof. I yelled up to my Dad that there was a large manure pile on the other side of the barn. My Dad asked me how far I had sunk jumping into the pile. I said, "I sank to my ankles." With the ladder now broken on the ground, my Dad walked to the other side and measured up the pile. Jumping in he sank up to his neck. My Dad said, "I thought you only sank to your ankles!" I replied "Yes, I jumped in head first."

SOME GROUND RULES FOR LIFE

- Keep moving. Exercise. Staying active at any age provides ample benefits, and that includes the ultimate natural exercise … gardening.

- Dig in. Go out there and kill some plants. If you haven't killed any plants you're not trying hard enough.

- Tell stories. The most useful stories are learned in the valley and not at the top of the hill. Be a "sonder" listening and sharing the stories of our amassed experiences and complex lives.

- Don't be a stick in the mud. If you insist, your lot in life may be chairman of the "bored."

- Be a gibberellin. Stretch yourself.

- Remember grace and unconditional love are like the water from a garden hose, it always flows to the lowest point. Find those at a low point and extend some grace. "No one is useless in this world that lightens the burden of another" -Charles Dickens

- Solid as a rock is not always good. Show some vulnerability and be a little "bolder."

- Entropy is a reality. You can't escape it, so learn to embrace it in a positive way to help others in the journey. Change is inevitable. You are either moving forward or backwards, but nothing ever stays the same.

- Sometimes you have to jump and figure out the landing on the way.

- You have to dig some holes to stand on mountains.

GROUND RULES FOR THE WEEKEND GARDENER

In order to establish lawn and order you need some ground rules as a weekend gardener and home owner. These general principles or rules will go a "lawn" way in helping you apply well-grounded basics around the home. (This top ten list pays for the cost of your book.)

Here is my top-ten list:

1. If you want to establish a great lawn and landscape, be out there in the fall season working on it. In September to November the soil is still warm, and we generally get sufficient rainfall and the air temperature cools. In fall, plants put their effort into root establishment, instead of top growth followed by the resting dormancy of winter. It's a great time of year to plant, move plants or split your plants. When you feed landscape plants in fall even though air temperatures are dropping, soil temperatures remain warm enough for roots to benefit from nutrients, giving them a jump-start on next year's growing season.

2. Take me to your weeder. Most people focus attention on weed killing in spring and summer. Spring is a good time of year for control of weed SEED, and applications of corn gluten, as an example, can help suppress weed seed hoping to germinate. Weeds, however, that are vegetative and visual can be very effectively controlled in **fall**. Winter annual weeds like henbit, that germinate and establish in fall and then appear and bloom seemingly out of nowhere in spring, are best controlled in fall. By the time they bloom in

spring, it's too late. They are going to die regardless. This makes your application a revenge spraying, since they have already produced copious amounts of seed to ensure you have to deal with it again next year. My arch nemesis, hairy bittercress, does this to me every spring. He poses as a nondescript innocent weed in fall, hides under the snow in winter, and when spring arrives pops up everywhere I look with its multitudinous little white blooms. Perennial weeds like dandelions or broadleaf plantain are more vulnerable in fall than spring, because, they, like the trees, are sending food reserves to the root system, shutting down for winter. If you apply weed controls at that time of the year, they translocate well into the root system, giving better control than the top kill we often get in spring. It's either that or accepting compliments on their robust growth and weeding by example by utilizing them for foraging and medicinal reasons

3. When it comes to the lawn, if you do nothing else, at least raise the deck on your lawn mower. Longer grass leaf blades shade the crown of the plant, reducing stress, provide more surface area for healthy photosynthesis and naturally help shade out weeds, giving the lawn a thicker more competitive edge.

4. Grow some vegetables in containers and eat more plants. Think sunshine, watering at the base when possible, and provide healthy organic-matter soil, because … remember, you are what you eat. That's why I avoid rump roast.

5. Pruning. If it blooms in spring prune right after blooming. Prune in June for evergreens. If it blooms in fall (August to October) prune in spring. Deadhead herbaceous perennials, and leave the foliage on, if it

looks nice as long as possible. Pinch mums and tall sedums until July 1. Non-flowering woody plants and trees prune in winter. With no leaves, insect or disease activity the dead of winter is the perfect time. Don't be afraid to rejuvenate-prune or prune to allow more sunlight penetration and air movement. Pruning produces outward growth, and can often stimulate more aggressive blooming (remember chapter 15?) With all that said ... **all rules have exceptions including the rule that says all rules have exceptions.** In other words, pruning will always cause duress and confusion for many homeowners, and can be the cause of marital disputes. So, the final word is this: pruning questions provide job security for someone like me.

6. Mulch. Mulch is a good thing, but like your Mom always told you, you can overdo a good thing. Mulch reduces soil temperature in summer and helps maintain a steady temperature the rest of the year. Mulch helps hold some moisture in the soil and suppresses light to weed seeds. Don't allow mulch to become hydrophobic, stir it occasionally. Don't pile it on year after year until you have a thick spongy surface. 1 to 2 inches is sufficient. Gases develop under "overpiled" mulch that can be toxic to your plants. Don't mound mulch against tree and woody plant trunks accustomed to being dry. Mulch "volcanoes" will cause girdling roots and crown rot at the base of the plant or tree. If you ordered too much mulch piled in your driveway in spring, share it with your neighbors. Your dogwood will say thank you very mulch.

7. Houseplants. Every home needs foliage plants; they improve our indoor air and mood. First, you need some bright-lit window areas. Don't kill with kind-

ness, as water is the number one killer of houseplants. Too much water rots roots, brings on decline and annoying fungus gnats. Where practical, lift a pot and tell by its weight if the plant needs water. Dust them and give an occasional quarter turn. Watch for bugs and honeydew, a sticky substance that is the tell-tale sign a pest is sucking the life out of your plants. Move them outside to a protected area in summer for a little vacation. Talk to your plants. They like the carbon dioxide. They would even enjoy some time in the bathroom from time to time if a shower is present. (But only if this isn't too embarrassing for you.) Don't worry, Phil-O-Dendron has seen it all. Tropical plants like humidity, not dry indoor air. Move them away from heat registers or drafty doorways. Houseplants don't like cold drafts, and we're not talking beer here. Grouping houseplants together helps too.

8. Water. There is NOT a one-size-fits-all watering schedule. (See all rules have exceptions rule in point 5 above, again ensuring my job security.) Most plants like a moist well-drained soil, which sounds like an oxymoron but it's true. Exposure and species is a factor. Time of year is a factor. Soil type is a factor. Invest in the foundation of a good quality soil, and it will pay for itself. Liberally work in organic matter 50/50 with the parent soil for outdoor plantings. Sprinklers are good for lawns, not for landscape plants. Trees and woody plants should have a hose-trickle, deep soaking at the base, from time to time, dependent on weather conditions. If you're really stressed about it, invest in a moisture meter. For hanging baskets, press against the base of the basket, lifting to feel the weight. Learn to tell by the weight of the basket if it needs water,

then water thoroughly. If soil dries, it contracts from the sides of the basket, and water flows across the top and down the sides. Take the basket down from time to time to drench the core. Water soluble polymer crystals are a good idea for container plantings. Moisture manager granules are available for landscape materials and lawns. Mycorrhizae is a wonderful natural way to colonize roots and enhance their reach into the soil.

9. Exposure. Think wind and sun. What can the species handle? Think of sun and wind exposure for all times of the year, especially winter. If the ground freezes, evergreens (like broadleaf evergreens) will desiccate in the winter sun and wind. The sun's position in winter, low on the south horizon, combined with wind, can dry out your plants unable to get a drink from frozen ground. That is why plants like broadleaf evergreens appreciate the structural protection of a home on the north or east side of the house in winter. Anti desiccant sprays of pine resin are also useful.

10. Above all else my friends, maintain your "composture" and your sense of "humus."

About the Author

Rick Vuyst is a self-made "Entre-manure" willing to get his hands dirty and put himself out there, like the fledgling spring blooms in a late-spring frost. Rick is CEO of *Flowerland* stores Grand Rapids, Michigan, with three garden centers serving the West Michigan area. Rick is host of the *Flowerland* show on NewsRadio WOOD 1300 and 106.9 FM, and nationally on IHeart radio as "Phil Dirt" "heducating" and entertaining audiences for 25 years. Rick has also been the "*Mr Green Thumb*" for WZZM TV 13, the West Michigan ABC television affiliate for over 20 years. Rick is a writer for *Women's Lifestyle Magazine* and a blogger on his personal site *thankyouverymulch.com*. In addition to gardening, Rick is a health enthusiast, avid runner and loves photography on the Lake Michigan shoreline and in the garden where he lives in West Michigan.

"Funny and incredibly smart – Rick Vuyst is the ultimate expert on anything plant-related. He makes hard topics easy to understand, and is always quick to add a perfect pun to any helpful tip."

-Ellen Bacca
WOOD TV 8 meteorologist

"Rick is a successful business owner and popular radio and TV personality, but he's never too busy to share his expertise and help anyone, doing so with patience, integrity and HUMOR. Being one of Rick's "partners in grime" has been a great privilege and I'm honored to call him a friend."

— Derek Francis,
News Anchor Fox 17
Grand Rapids, MI

"Rick's passion stems beyond plants and running. His willingness to help a friend might be one of his most endearing qualities. Rick truly lives his life in full bloom, and I cherish being able to benefit from his knowledge, talent and kindness."

— Valerie Lego, Health Reporter,
WZZM 13 Grand Rapids

"Have you ever heard Rick Vuyst on WOOD Radio and wished he could just come to your house and help? Well that's what this book is really ... him coming to your house to help ... Without you feeling guilty about taking him away from his family. And he'll have more time to think up puns this way."

— Steve Kelly, fan and Host
of West Michigan's Morning News
Newsradio WOOD 106.9 FM 1300 AM

"Rick has all sorts of secrets for successful gardening, and he shares his knowledge with a fun sense of humor. His gardening tips have even made me a green thumb."

-Janet Mason President & General Manager
WZZM 13 Grand Rapids, Michigan

"Rick is very dedicated to his plants. He once stood still for an entire month just to better understand his magnolia tree."

— Steve Zaagman National Board Game
Champion and TV personality

"I have known Rick for more than a decade as a truly passionate, disciplined leader with an insatiable appetite for learning. His whip-smart mind makes him a formidable (and occasionally groan-inducing) punster, able to coax a smile out of any crowd. His lens on the world (both literally and metaphorically) never fails to provide me with an interesting perspective, and I always look forward to his insight on any given situation. It's always an honor - and a pleasure - to work with him!"

— Kellee O'Reilly,
Chief Experience Officer
MonkeyBar Management

"Rick Vuyst is this wonderful blend of scientist and storyteller who has shared no-nonsense advice with West Michigan television viewers and radio listeners going on three decades now. I am proud to have worked with Rick for most of those years and I now share his advice like it's my own. I promise that this book is a clever blend of humor, nostalgia, and good advice, whether or not you ever step foot in a garden. I'm proud to have worked with Rick for most of those years and cannot wait to dive into some of his favorite stories, "book, line, and sinker!"

— Catherine Behrendt
News Anchor and Television Host
WZZM 13 – Grand Rapids, Michigan

"I've had the pleasure of working with Rick for over 25 years and it's been a joy watching him "cultivate his talents" on WOOD Radio. Rick is a very rare and uniquely talented radio host...genuinely authentic, kind and thoughtful, and never without a pun or phrase that delights and entertains his listeners. "I just Wet My Plants" is the perfect "next thing" for my favorite entremanure!"

— Phil Tower Program Director
Newsradio WOOD 106.9 FM 1300 AM

"For years, my relationship with Rick was based on our broadcasting work, but that all changed during a "not-so-great" season in my personal life. Rick's take on my period of "entropy" (see chapters 7 and 9) was incredibly insightful, uplifting and comforting. He's one of those people who makes you feel good about life's journey, just by being in his presence."

— Jennifer Pasqua WZZM TV News
and Co-host for *My West Michigan*

"Every gardener and wanna-be gardener should read this amusing book. Rick Vuyst goes beyond expert gardening advice to bring us entertaining anecdotes of the pleasures and pains of keeping up the back yard."

— Pam Spring
Pam Spring Advertising Agency

"We aren't all blessed to have been born with a green thumb, so thank goodness Rick Vuyst is as gifted with a microphone as he is with a shovel! Rick's special wit and vast wisdom have certainly produced more than a bumper crop of great growing seasons. Though his taste in ice cream creations remains a mystery." (See chapter 4 for details.)

— Terri DeBoer WOOD TV 8
Meteorologist Grand Rapids Michigan

"It's been said that a pun is the lowest form of comedy. But with Rick, it not only takes a higher form, it is usually wrapped in some knowledge and enthusiasm that makes me want to make my garden, yard, and world a better place"

— Tracy Forner,
TV Host/Inspired Gardener

"Honored to be your friend in flora from the wizard of weather!"

— Chief Meteorologist
George Lessens WZZM TV 13

"This is not a book about being a great gardener, in fact it is about the process of becoming a joyful gardener. Rick Vuyst has spent a good deal of his life amassing stories, ideas and more importantly, experiences. He shares those freely and allows us to participate in his journeys, his insights and his love of this thing we call gardening. Gardening is not brain surgery, and he encourages to have fun and to experiment. Thank you Rick, I am killing more plants as we speak."

— Dr. Allan Armitage
Professor Emeritus of Horticulture
University of Georgia